Fundamentals of Structural Engineering

Fundamentals of Structural Engineering

Daniel Campbell

Willford Press
www.willfordpress.com

Published by Willford Press,
118-35 Queens Blvd., Suite 400,
Forest Hills, NY 11375, USA

ISBN: 978-1-64728-325-4

Cataloging-in-Publication Data

Fundamentals of structural engineering / Daniel Campbell.
p. cm.
Includes bibliographical references and index.
ISBN 978-1-64728-325-4
1. Structural engineering. 2. Engineering. 3. Architecture. I. Campbell, Daniel.
TA633 .F86 2022
624.1--dc23

For information on all Willford Press publications
visit our website at www.willfordpress.com

WILLFORD PRESS

Contents

Preface

The branch of engineering which focuses on the design and analysis of structures that support or resist loads is known as structural engineering. The theory of structural engineering is based on the laws of applied physics and the empirical knowledge of structural performance of different materials and geometries. There are various specializations within this field such as earthquake engineering structures, mechanical structures, aerospace structures, and civil engineering structures. Earthquake engineering structures deal with the construction of structures which are specially engineered to withstand earthquakes. Some of the numerous elements, used in structural engineering are arches, columns, shells, beams and plates. This book brings forth some of the most innovative concepts and elucidates the unexplored aspects of structural engineering. It will serve as a valuable source of reference for graduate and postgraduate students. Coherent flow of topics, student-friendly language and extensive use of examples make this book an invaluable source of knowledge.

A foreword of all chapters of the book is provided below:

Chapter 1 - Structural engineering is the discipline of civil engineering which deals with the analysis, planning and designing of modern buildings. Structural engineers calculate the stability, strength and rigidity of structural designs. This chapter has been carefully written to provide an easy understanding of structural engineering; **Chapter 2 -** Structural elements are used in structural analysis for splitting of a complex structure into simple elements. It includes beams, trusses, columns, cables, arches, etc. The topics elaborated in this chapter will help in gaining a better perspective about these structural elements; **Chapter 3 -** Structural materials are mainly used for their mechanical properties. These materials include iron, steel, aluminium, timber, bamboo, carbon fibre, adobe, fibre reinforced polymer, etc. This chapter closely examines these different structural materials to provide an extensive understanding of the subject; **Chapter 4 -** Structural stability aims at ensuring the safety of structures against failures in engineering designs and plans. The basic types of structural failure include compressive, tensile, bending and buckling. This chapter discusses about the aspects of structural stability and failure in detail; **Chapter 5 -** Foundations can be defined as the element that connects the structure to the ground and helps in transferring the load from the ground. Deep foundations and shallow foundations are the two types that fall under its domain. The diverse applications of foundations in structural engineering have been thoroughly discussed in this chapter.

At the end, I would like to thank all the people associated with this book devoting their precious time and providing their valuable contributions to this book. I would also like to express my gratitude to my fellow colleagues who encouraged me throughout the process.

Daniel Campbell

1

Introduction to Structural Engineering

Structural engineering is the discipline of civil engineering which deals with the analysis, planning and designing of modern buildings. Structural engineers calculate the stability, strength and rigidity of structural designs. This chapter has been carefully written to provide an easy understanding of structural engineering.

Structure

Within the context of the built environment, the term 'structure' refers to anything that is constructed or built from different interrelated parts with a fixed location on the ground.

This includes buildings, but the term structure can also be used to refer to anybody of connected parts that is designed to bear loads, even if it is not intended to be occupied by people. Engineers sometimes refer to these as 'non-building' structures. Common examples include:

- Aqueducts and viaducts,
- Bridges,
- Canals,
- Cooling towers and chimneys,
- Dams,
- Railways,
- Roads,
- Retaining walls,
- Tunnels.

Components of a Building Structure

The basic components of a building structure are the foundation, floors, walls, beams, columns, roof, stair, etc. These elements serve the purpose of supporting, enclosing and protecting the building structure.

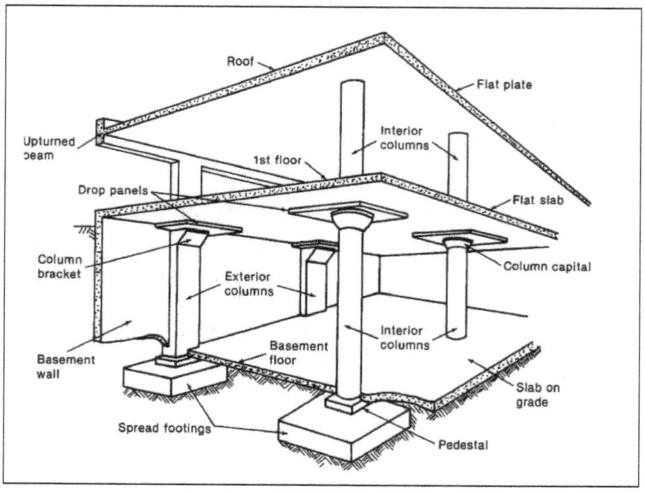

Basic Components of a Building.

Mentioned below are the 12 basic components a building structure:

- Roof: The roof forms the topmost component of a building structure. It covers the top face of the building. Roofs can be either flat or sloped based on the location and weather conditions of the area.

- Parapet: Parapets are short walls extended above the roof slab. Parapets are installed for flat roofs. It acts as a safety wall for people using the roof.

- Lintels: Lintels are constructed above the wall openings like doors, windows, etc. These structures support the weight of the wall coming over the opening. Normally, lintels are constructed by reinforced cement concrete. In residential buildings, lintels can be either constructed from concrete or from bricks.

- Beams and slabs: Beams and slabs form the horizontal members in a building. For a single storey building, the top slab forms the roof. In case of a multi-storey building, the beam transfers the load coming from the floor above the slab which is in turn transferred to the columns. Beams and slabs are constructed by reinforced cement concrete (R.C.C.).

- Columns: Columns are vertical members constructed above the ground level. Columns can be of two types: Architectural columns and structural columns. Architectural columns are constructed to improve the building's aesthetics while a structural column takes the load coming from the slab above and transfers safely to the foundation.

- Damp Proof Course (DPC): DPC is a layer of waterproofing material applied on the basement level to prevent the rise of surface water into the walls. The walls are constructed over the DPC.

- Walls: Walls are vertical elements which support the roof. It can be made from stones, bricks, concrete blocks, etc. Walls provide an enclosure and protect against wind, sunshine, rain etc. Openings are provided in the walls for ventilation and access to the building.
- Floors: The floor is the surface laid on the plinth level. Flooring can be done by a variety of materials like tiles, granites, marbles, concrete, etc. Before flooring, the ground has to be properly compacted and leveled.
- Stairs: A stair is a sequence of steps that connects different floors in a building structure. The space occupied by a stair is called as the stairway. There are different types of stairs like a wooden stair, R.C.C stair etc.
- Plinth Beam: Plinth beam is a beam structure constructed either at or above the ground level to take up the load of the wall coming over it.
- Plinth: The plinth is constructed above the ground level. It is a cement-mortar layer lying between the substructure and the superstructure.
- Foundation: The Foundation is a structural unit that uniformly distributes the load from the superstructure to the underlying soil. This is the first structural unit to be constructed for any building construction. A good foundation prevents settlement of the building.

Types of Structures

Trusses

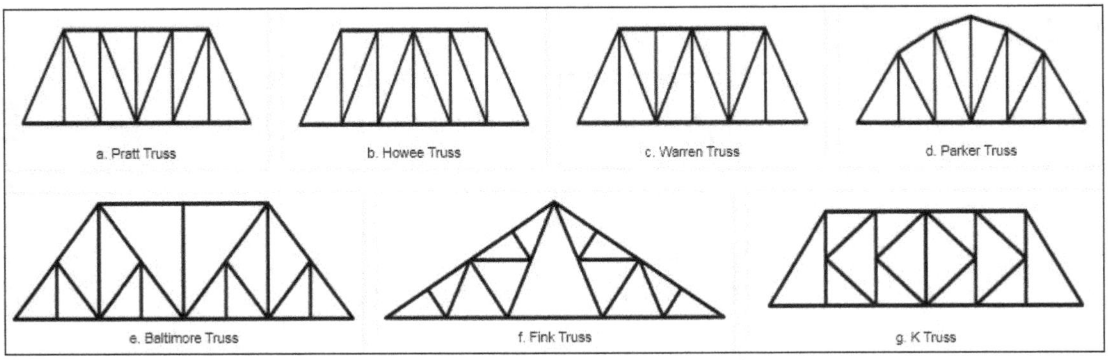

Types of Plane Trusses.

As spans increase the use of beams becomes uneconomical. For moderately long span trusses are commonly used. Trusses consist of slender elements connected at their ends (joints) by hinged connections arranged in triangular fashion to form a stable configuration. When the loads are applied at the joints, ideal trusses are always either in uniform tension cause member elongate or in uniform compression cause member shorten. Real trusses, although some bending formed in the members when it is loaded, in most cases such secondary stresses are small, and the assumption of pure tension and compression are satisfactory. Trusses are commonly used in bridge and roof structure. Based on their shape, trusses are classified into several types shown in figure.

Frames

Frames are composed of beams and columns that are either rigid (moment-resisting) connection or by hinged connection to form stable configurations. Unlike trusses, which are subjected only to joint loads, the external loads on frames may be applied on the members as well as on the joints. The members of frame subjected to bending moment, shear, axial (compression or tension), torsion (for space trusses) under the action of external loads. Structural steel and reinforced concrete frames are commonly used in multi-storey buildings and industrial plants. Example shape of frame shown in figure.

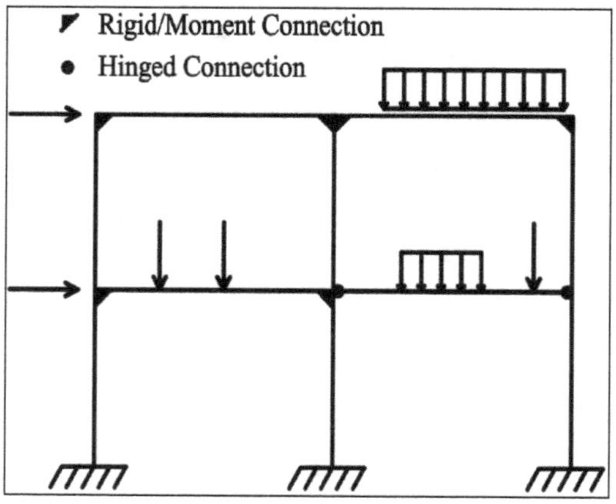

Frame Structure.

Arches

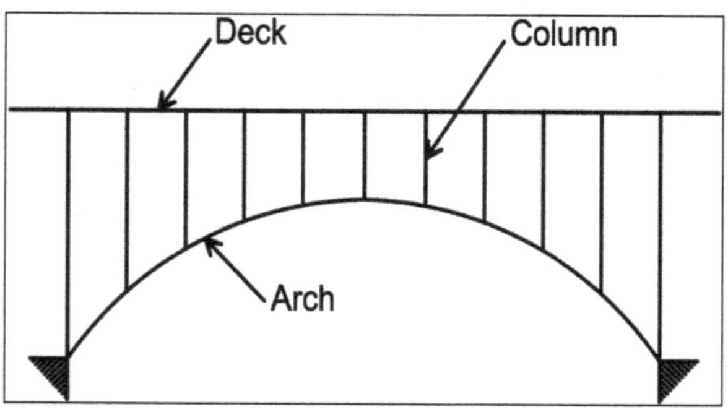

Arch Structure: Bridge Deck Supported by Column.

The use of trusses to support bridge deck becomes impracticable for longer than moderate spans. In this situation arches are commonly used. Figure shows an arch in which the bridge deck is carried by columns supported, in turn, by the arch.

Alternatively the bridge deck may be suspended from the arch by hangers as shown in figure. Arches carry most of their loads by developing compressive stresses within the arch itself.

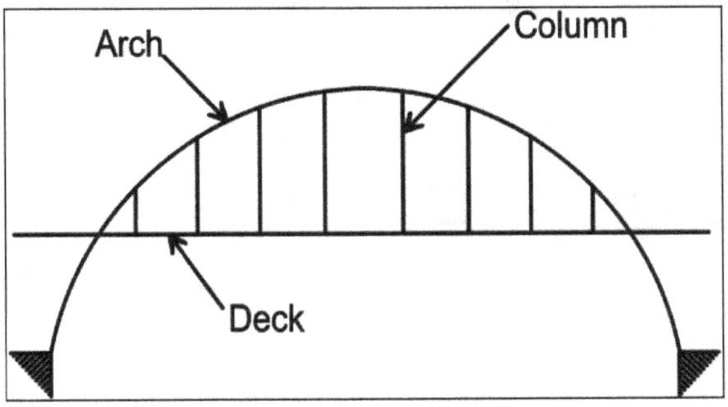

Arch Structure: Bridge Deck Supported by Hanger.

Cables

Other structures than arches to support long-span bridges are cables. Because of their flexibility, cables have negligible bending stiffness and can develop only tension. Thus, under external loads, a cable adopts a shape that enables it to support the load by tensile force alone. In other words, the shape of a cable changes as the loads acting on it change. One of the popular cable structures is suspension bridge. In suspension bridge, the roadway suspended from two main cables by means of vertical hangers. The cables pass over saddles on the tops of tower and are fixed at each end within the ground by massive anchor blocks. The weakness of suspension bridge and other cable structure is lack of stiffness in lateral directions; they are susceptible to wind-induced oscillations. Bracing of stiffening system are therefore provided to reduce such oscillations.

Structural Idealization

To calculate the forces in the various parts of a structure with reasonable simplicity and accuracy, it is necessary to represent the structure in a simple manner that is conducive to analysis. Structural components have width and thickness. Concentrated forces rarely act at a single point; rather, they are distributed over small areas. If these characteristics are taken into consideration in detail, however, an analysis of the structure will be very difficult, if not impossible to perform.

The process of replacing an actual structure with a simple system conducive to analysis is called structural idealization. Most often, lines that are located along the centrelines of the components represent the structural components. The sketch of a structure idealized in this manner usually is called a line diagram.

The preparation of line diagrams is shown in figure. The wood beam shown supports several floor joists and in turn is supported by three concrete block walls. The actual distribution of the forces acting on the beam is also shown. For purposes of analysis, though, we can conservatively represent the beam and its loads and reactions with the line diagram. The loaded spans are longer with the result that shears and moments are higher than actually occur.

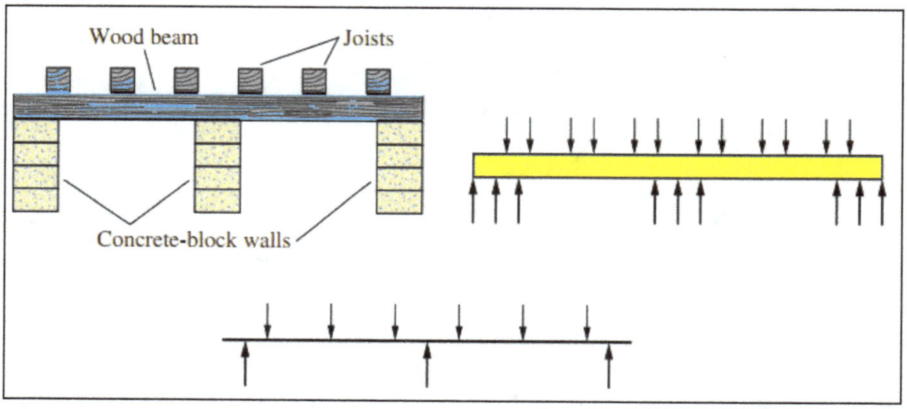

Sometimes the idealization of a structure involves assumptions about the behavior of the structure. As an example, the bolted steel roof truss of figure is considered. The joints in trusses often are made with large connection or gusset plates and, as such, can transfer moments to the ends of the members. However, experience has shown that the stresses caused by the axial forces in the members greatly exceed the stresses caused by flexural forces. As a result, for purposes of analysis we can assume that the truss consists of a set of pin-connected lines.

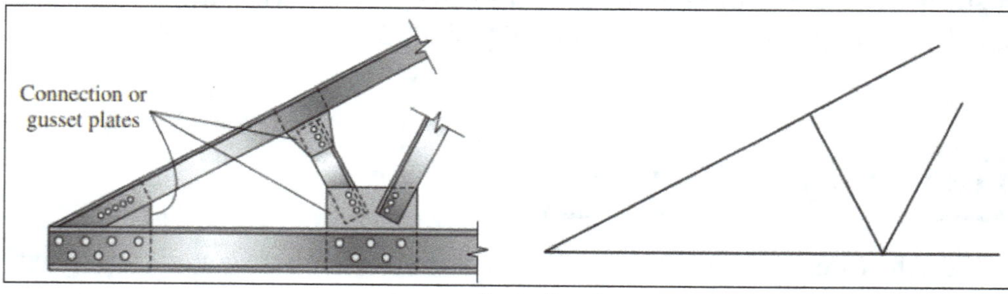

Structural Idealization.

Although the use of simple line diagrams for analyzing structures will not result in perfect analyses, the results usually are quite acceptable. Sometimes, though, there may be some doubt in the mind of the analyst as to the exact line diagram or model to be used for analyzing a particular structure. For instance, should beam lengths be clear spans between supports, or should they equal the distances center to center of those supports? Should it be assumed that the supports are free to rotate under loads, are fixed against rotation, or do they fall somewhere in between? Because of many questions such as these, it may be necessary to consider different models and perform the analysis for each one to determine the worst cases.

Structural Design

Structural design is the methodical investigation of the stability, strength and rigidity of structures. The basic objective in structural analysis and design is to produce a structure capable of resisting all applied loads without failure during its intended life. The primary purpose of a structure is to transmit or support loads. If the structure is improperly designed or fabricated, or if the actual applied loads exceed the design specifications, the device will probably fail to perform its intended function, with possible serious consequences. A well- engineered structure greatly minimizes the possibility of costly failures.

Structural Design Process

A structural design project may be divided into three phases, i.e. planning, design and construction.

- Planning: This phase involves consideration of the various requirements and factors affecting the general layout and dimensions of the structure and results in the choice of one or perhaps several alternative types of structure, which offer the best general solution. The primary consideration is the function of the structure. Secondary considerations such as aesthetics, sociology, law, economics and the environment may also be taken into account. In addition there are structural and constructional requirements and limitations, which may affect the type of structure to be designed.

- Design: This phase involves a detailed consideration of the alternative solutions defined in the planning phase and results in the determination of the most suitable proportions, dimensions and details of the structural elements and connections for constructing each alternative structural arrangement being considered.

- Construction: This phase involves mobilization of personnel; procurement of materials and equipment, including their transportation to the site, and actual on-site erection. During this phase, some redesign may be required if unforeseen difficulties occur, such as unavailability of specified materials or foundation problems.

Philosophy of Designing

The structural design of any structure first involves establishing the loading and other design conditions, which must be supported by the structure and therefore must be considered in its design. This is followed by the analysis and computation of internal gross forces, (i.e. thrust, shear, bending moments and twisting moments), as well as stress intensities, strain, deflection and reactions produced by loads, changes in temperature, shrinkage, creep and other design conditions. Finally comes the proportioning and selection of materials for the members and connections to respond adequately to the effects produced by the design conditions.

The criteria used to judge whether particular proportions will result in the desired behavior reflect accumulated knowledge based on field and model tests, and practical experience. Intuition and judgment are also important to this process.

The traditional basis of design called elastic design is based on allowable stress intensities which are chosen in accordance with the concept that stress or strain corresponds to the yield point of the material and should not be exceeded at the most highly stressed points of the structure, the selection of failure due to fatigue, buckling or brittle fracture or by consideration of the permissible deflection of the structure. The allowable – stress method has the important disadvantage in that it does not provide a uniform overload capacity for all parts and all types of structures.

The newer approach of design is called the strength design in reinforced concrete literature and plastic design in steel-design literature. The anticipated service loading is first multiplied by a suitable load factor, the magnitude of which depends upon uncertainty of the loading, the possibility of it changing during the life of the structure and for a combination of loadings, the likelihood, frequency, and duration of the particular combination. In this approach for reinforced-concrete design, theoretical capacity of a structural element is reduced by a capacityreduction factor to provide for small adverse variations in material strengths, workmanship and dimensions. The structure is then proportioned so that depending on the governing conditions, the increased load cause fatigue or buckling or a brittle-facture or just produce yielding at one internal section or sections or cause elastic-plastic displacement of the structure or cause the entire structure to be on the point of collapse.

Design Aids

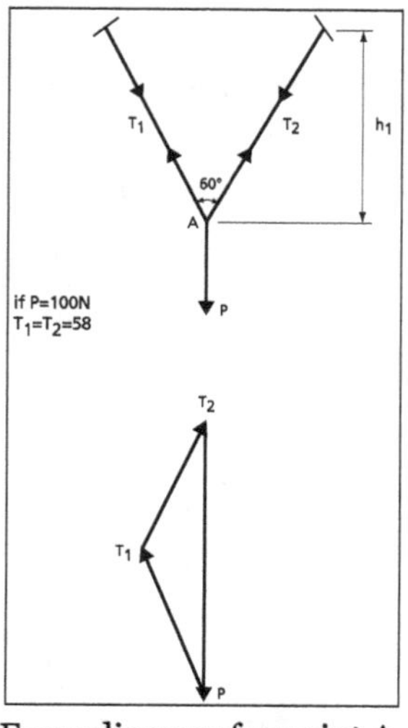

Force diagram for point A.

The design of any structure requires many detailed computations. Some of these are of a routine nature. An example is the computation of allowable bending moments for standard sized, species and grades of dimension timber. The rapid development of the computer in the last decade has resulted in rapid adoption of Computer Structural Design Software that has now replaced the manual computation. This has greatly reduced the complexity of the analysis and design process as well as reducing the amount of time required to finish a project.

Standard construction and assembly methods have evolved through experience and need for uniformity in the construction industry.

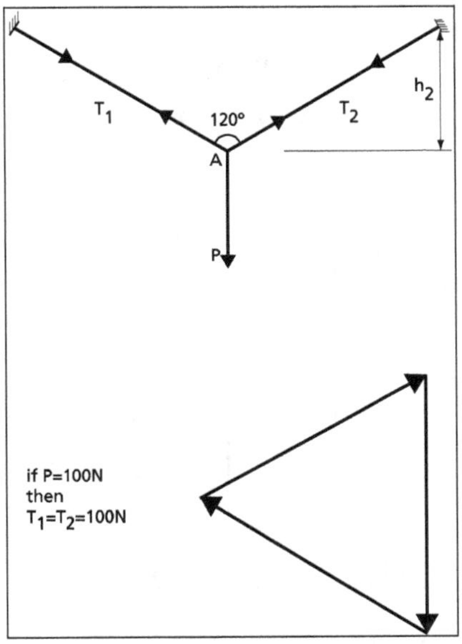

Force diagram for point A.

Design Codes

Many countries have their own structural design codes, codes of practice or technical documents which perform a similar function. It is necessary for a designer to become familiar with local requirements or recommendations in regard to correct practice.

Design of Members in Direct Tension and Compression

Tensile Systems

Tensile systems allow maximum use of the material because every fibre of the cross-section can be extended to resist the applied loads up to any allowable stress. As with other structural systems, tensile systems require depth to transfer loads economically across a span. As the sag (h) decreases, the tensions in the cable (T_1 and T_2) increase. Further decreases in the sag would again increase the magnitudes of T_1 and T_2 until the ultimate condition, an infinite force, would be required to transfer a vertical load across a cable that is horizontal (obviously an impossibility).

A distinguishing feature of tensile systems is that vertical loads produce both vertical and horizontal reactions. As cables cannot resist bending or shear, they transfer all loads in tension along their lengths. The connection of a cable to its supports acts as a pin joint (hinge), with the result that the reaction (*R*) must be exactly equal and opposite to the tension in the cable (*T*). The *R* can be resolved into the vertical and horizontal directions producing the forces *V* and *H*. The horizontal reaction (*H*) is known as the thrust.

The values of the components of the reactions can be obtained by using the conditions of static equilibrium and resolving the cable tensions into vertical and horizontal components at the support points.

Example: Two identical ropes support a load *P* of 5 kN, as shown in the figure. Calculate the required diameter of the rope, if its ultimate strength is 30 MPa and a safety factor of 4.0 is applied. Also determine the horizontal support reaction at *B*.

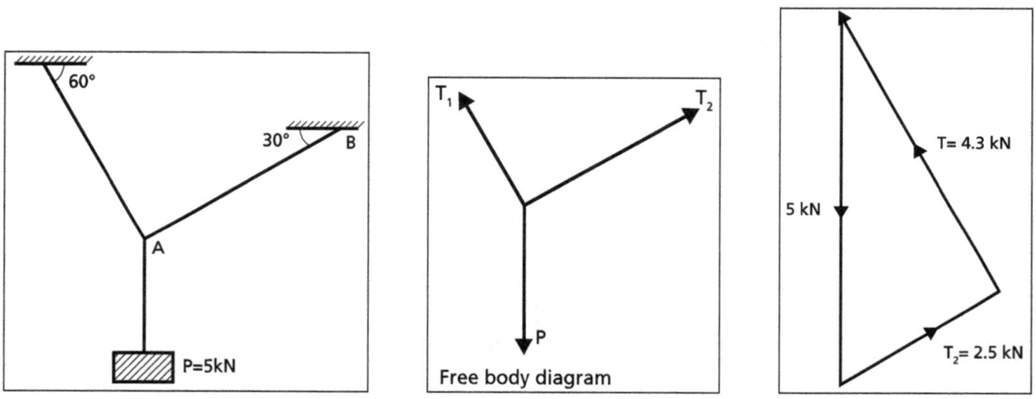

The allowable stress in the rope is:

$$\frac{30}{4} = 7.5\ N/mm^2 = 7.5\,MPa$$

$$\text{Stress} = \frac{\text{Force}}{\text{Area}}$$

Therefore:

$$\text{Area required} = \frac{4.3\times10^3}{7.5} = 573\ mm^3$$

$$A = \pi r^2 = \frac{\pi d^2}{4}$$

Thus,

$$d = \sqrt{\frac{4\times573}{\pi}} = 27\,mm\,(min)$$

At support B, the reaction is composed of two components:

Bv = T2 sin 30° = 2.5 sin 30° = 1.25 kN

BH = T2 cos 30° = 2.5 cos 30° = 2.17 kN

Short Columns

A column which is short (i.e. the height is small compared with the cross-section area) is likely to fail because of crushing of the material.

However, that slender columns, which are tall compared with the cross-section area, are more likely to fail from buckling under a load much smaller than that needed to cause failure from crushing. Buckling is dealt with later.

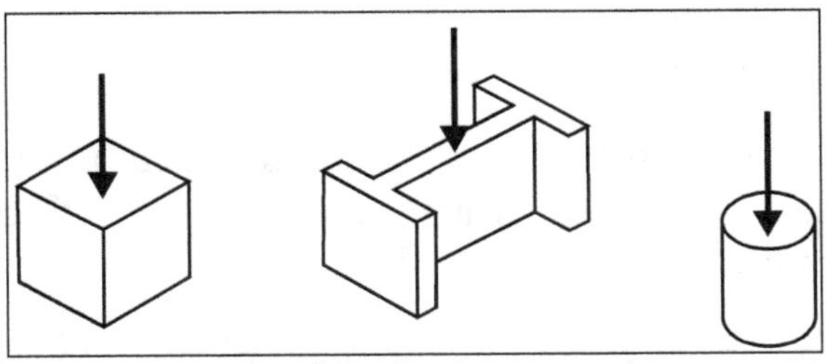

Short Columns

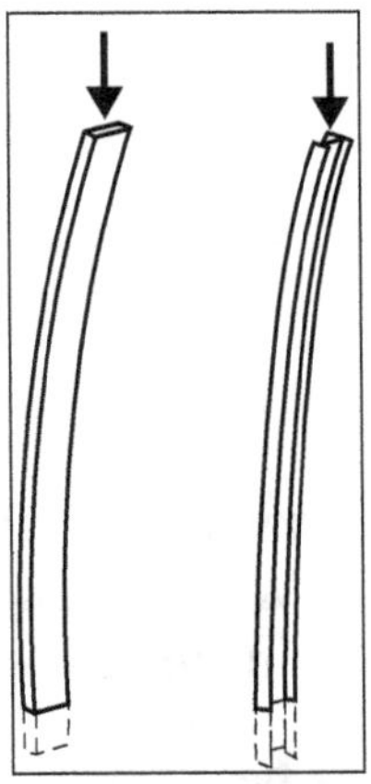

Slender Columns

Example: A square concrete column, which is 0.5 m high, is made of a nominal concrete mix of 1:2:4, with a permissible direct compression stress of 5.3 MPa (N / mm2). What is the required cross-section area if the column is required to carry an axial load of 300 kN?

$$A = \frac{F}{\sigma} = \frac{300000\,\text{N}}{5.3\ \text{N/mm}^2} = 56.604\ \text{mm}^2$$

i.e. the column should be minimum 238 mm square.

Design of Simple Beams

Bending Stresses

When a sponge is put across two supports and gently pressed downwards between the supports, the pores at the top will close, indicating compression, and the pores at the bottom will open wider, indicating tension. Similarly, a beam of any elastic material, such as wood or steel, will produce a change in shape when external loads are acting on it.

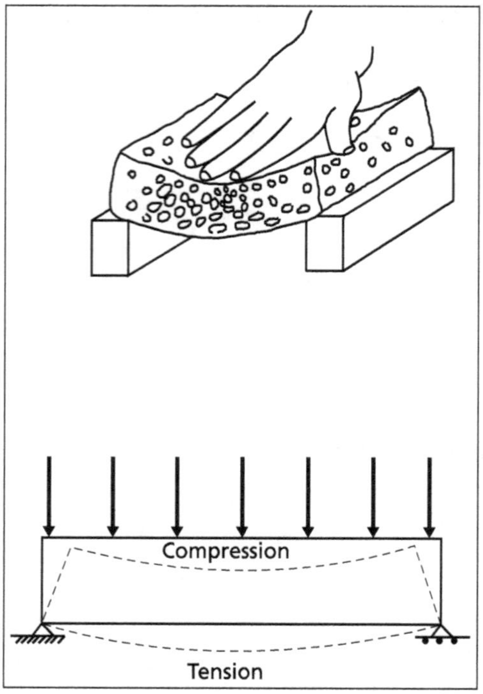

Bending effects on beams.

The stresses will vary from maximum compression at the top to maximum tension at the bottom. Where the stress changes from compressive to tensile, there will be one layer that remains unstressed and this is called the neutral layer or the neutral axis (NA).

This is why beams with an I-section are so effective. The main part of the material is concentrated in the flanges, away from the neutral axis. Hence, the maximum stresses occur where there is maximum material to resist them.

If the material is assumed to be elastic, then the stress distribution can be represented by two triangular shapes with the line of action of the resultant force of each triangle of stress at its centroid.

The couple produced by the compression and tension triangles of stress is the internal-reaction couple of the beam section.

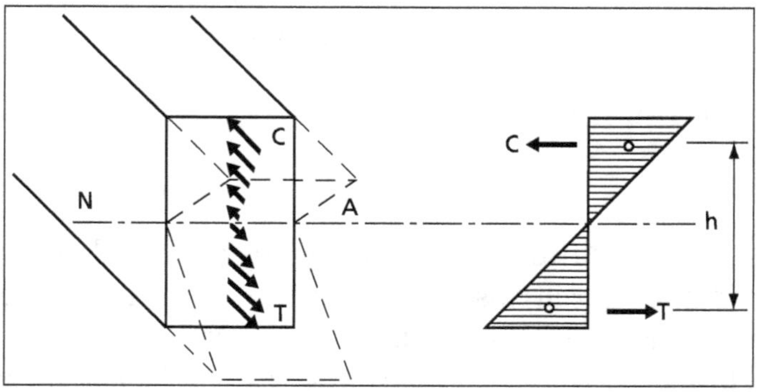

The moment caused by the external loads acting on the beam will be resisted by the moment of this internal couple. Therefore:

$$M = M_R = C(\text{or}\, T) \times h$$

Where,

- M = the external moment.
- M_R = the internal resisting moment.
- C = resultant of all compressive forces on the cross-section of the beam.
- T = resultant of all tensile forces on the cross-section of the beam.
- h = lever arm of the reaction couple.

Now consider a small element with the area (R) at a distance (a) from the neutral axis (NA).

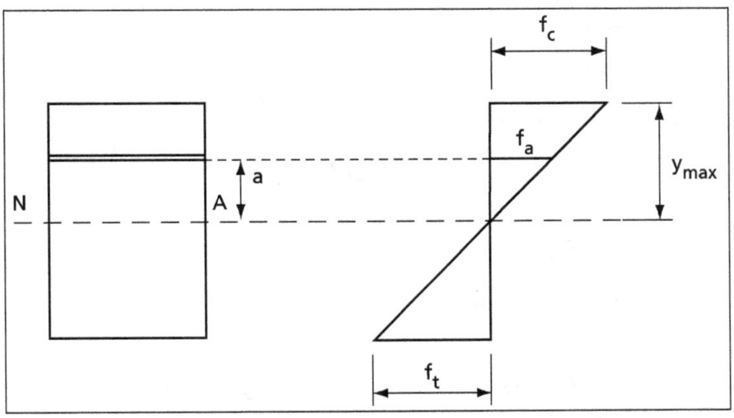

Note that it is common practice to use the symbol f for bending stress, rather than the more general symbol. Maximum compressive stress (f_c) is assumed to occur in this case at the top of the beam. Therefore, by similar triangles, the stress in the chosen element is:

$$\frac{f_a}{a} = \frac{f_c}{y_{max}},\ f_a = a \times \frac{f_c}{y_{max}}$$

As force = stress × area, then the force on the element = $f_a \times R = a \times (f_c / y_{max}) \times R$

The resisting moment of the small element is: force × distance:

$$(a) = a \times f_c / y_{max} \times R \times a = Ra^2 \times (f_c / y_{max})$$

The total resisting moment of all such small elements in the cross-section is:

$$M_R = \Sigma\, Ra^2 \times (f_c / y_{max})$$

But, $\sum Ra^2 = I$, the moment of inertia about the neutral axis, and therefore:

$$M_R = I \times (f_c / y_{max})$$

As the section modulus $Z_c = I/y_{max}$ therefore:

$$M_R = f_c \times Z_c = M;$$

Similarly,

$$M_R = f_t \times Z_t = M$$

The maximum compressive stress (f_c) will occur in the cross-section area of the beam where the bending moment (M) is greatest. A size and shape of cross-section, i.e. its section modulus (Z), must be selected so that the fc does not exceed an allowable value. As the following diagrams show, the concept of a 'resisting' couple can be seen in many structural members and systems.

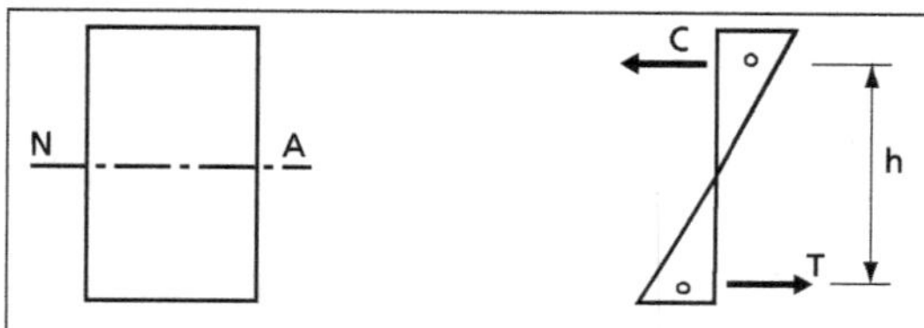

Rectangular beams

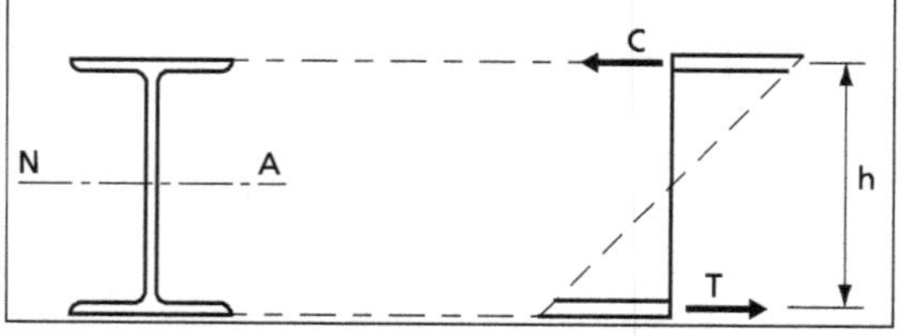

Girders and I –beams (1 /6 web area can be added to each flange area for moment resistance).

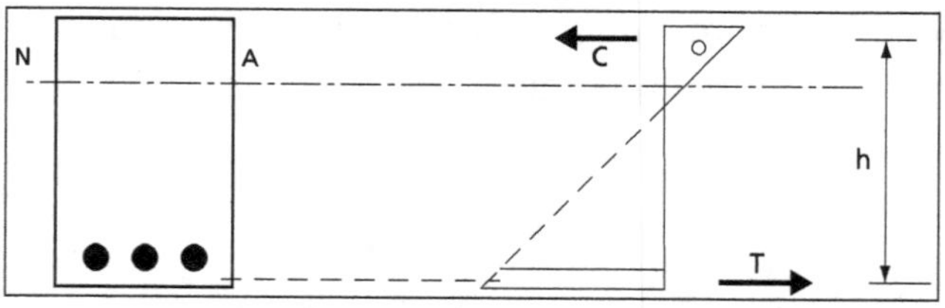

Rectangular reinforced-concrete beams (note that the steel bars are assumed to carry all the tensile forces).

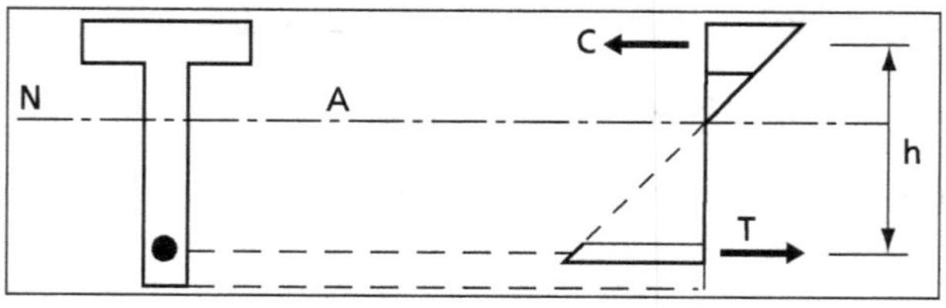

Reinforced-concrete T-beams.

The following equation is used to test for safe bending:

$$f_w \geq f = M_{max}/Z$$

Where,

- f_w = allowable bending stress,
- f = actual bending stress,
- M_{max} = maximum bending moment,
- Z = section modulus.

Horizontal Shear

The horizontal shear force (Q) at a given cross-section in a beam induces a shearing stress that acts tangentially to the horizontal cross-sectional plane. The average value of this shear stress is:

$$\tau = \frac{Q}{A}$$

Where, A is the transverse cross-sectional area.

This average value is used when designing rivets, bolts and welded joints. The existence of such a horizontal stress can be illustrated by bending a paper pad. The papers will slide relative to each other, but in a beam this is prevented by the developed shear stress.

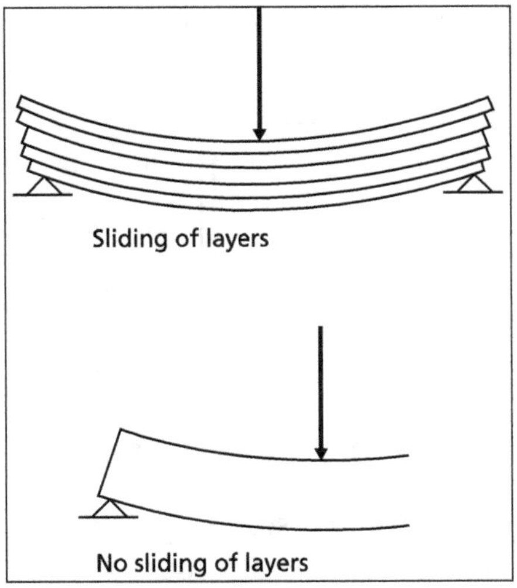

Shearing effects on beams.

However, the shear stresses are not equal across the cross-section. At the top and bottom edge of the beam they must be zero, because no horizontal shear stresses can develop.

If the shear stresses at a certain distance from the neutral axis are considered, their value can be determined according to the following formula:

$$\tau = \frac{Q \times \Delta A \times \overline{y}}{I \times b}$$

Where,

- t = shear stress,
- Q = shear force,
- ΔA = area for the part of the section being sheared off,
- $\overline{y}$ = perpendicular distance from the centroid of PA to the neutral axis,
- I = moment of inertia for the entire cross-section,
- b = width of the section at the place where shear stress is being calculated.

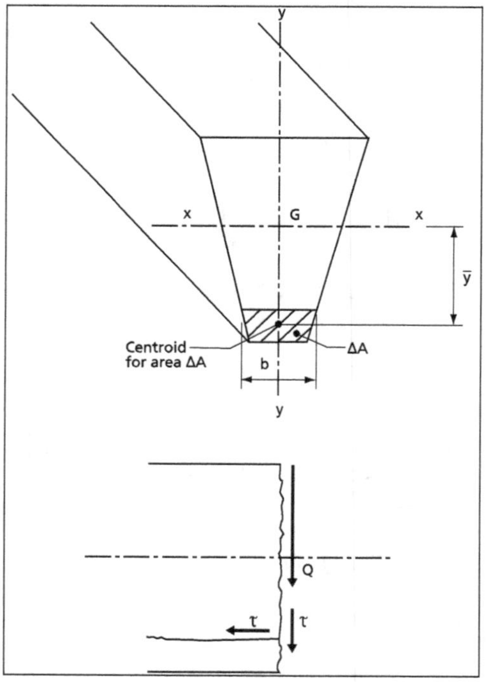

Maximum Horizontal Shear Force in Beams

It can be shown that the maximum shear stress τ_{max} in a beam will occur at the neutral axis. Thus, the following relations for the maximum shear stress in beams of different shapes can be deduced, assuming the maximum shear force (Q) to be the end reaction at a beam support (column).

For rectangular sections $\tau_{max} = \frac{3Q}{2bd} = \frac{3Q}{2A} = 1.5\frac{Q}{A}$

For square sections $\tau_{max} = \dfrac{3Q}{2a^2} = 1.5\dfrac{Q}{A}$

For circular sections $\tau_{max} = \dfrac{16Q}{3\pi D^2} = \dfrac{4Q}{3A}$

For I-shaped sections of steel beams, a convenient approximation is to assume that all shearing resistance is afforded by the web plus the part of the flanges that forms a continuation of the web.

Thus,

For I-sections $\tau_{max} \approx \dfrac{Q}{d \times t}$

Where,

- d = depth of beam,
- t = thickness of web.

If timber and steel beams with spans normally used in buildings are made large enough to resist the tensile and compressive stresses caused by bending, they are usually strong enough to resist the horizontal shear stresses also. However, the size or strength of short, heavily loaded timber beams may be limited by these stresses.

Deflection of beams Excessive deflections are unacceptable in building construction, as they can cause cracking of plaster in ceilings and can result in jamming of doors and windows. Most building codes limit the amount of allowable deflection as a proportion of the member's length, i.e. 1 /180, 1 /240 or 1 /360 of the length.

For standard cases of loading, the deflection formulae can be expressed as:

$$\delta_{max} = K_c \times \frac{WL^3}{EI}$$

Where,

- δ_{max} = maximum deflection (mm),
- K_c = constant depending on the type of loading and the end support conditions,
- W = total load (N),
- L = effective span (mm),
- E = modulus of elasticity (N/mm^2),
- I = moment of inertia (mm^4).

It can be seen that deflection is greatly influenced by the span L, and that the best resistance is provided by beams which have the most depth (d), resulting in a large moment of inertia.

Note that the effective span is greater than the clear span. It is convenient to use the centre to centre distance of the supports as an approximation of the effective span.

Design Criteria

The design of beams is dependent upon the following factors:

- Magnitude and type of loading,
- Duration of loading,
- Clear span,
- Material of the beam,
- Shape of the beam cross-section.

Beams are designed using the following formulae:

1. Bending stress:

$$f_w \geq f = \frac{M_{max}}{Z}$$

Where,

- f_w = allowable bending stress,
- f = actual bending stress,
- M_{max} = maximum bending moment,
- Z = section modulus.

This relationship derives from simple beam theory:

$$\frac{M_{max}}{I_{NA}} \quad \frac{f_{max}}{y_{max}}$$

and

$$\frac{I_{NA}}{y_{max}} = Z$$

The maximum bending stress will be found in the section of the beam where the maximum bending moment occurs.

2. Shear stress:

For rectangular cross-sections:

$$\tau_w \geq \tau = \frac{3 \times Q_{max}}{2 \times A} = \frac{3Q_{max}}{2bd}$$

For circular cross-sections:

$$\tau_w \geq \tau = \frac{4 \times Q_{max}}{3 \times A} = \frac{16Q_{max}}{3\pi d^2}$$

For I-shaped cross-sections of steel beams:

$$\tau_w \geq \tau = \frac{Q_{max}}{A}$$

Where,

- τ_w = allowable shear stress
- τ = actual shear stress
- Q_{max} = maximum shear force
- A = cross-section area

Like allowable bending stress, allowable shear stress varies for different materials and can be obtained from a building code.

3. Deflection:

In addition, limitations are sometimes placed on maximum deflection of the beam (δ_{max}):

$$\delta_{max} = K_c \times \frac{WL^3}{EI}$$

Example: Consider a floor where beams are spaced at 1 200 mm and have a span of 4 000 mm. The beams are seasoned cypress with the following properties:

f_w = 8.0 N/mm2, τ_w = 0.7 MPa (N/mm2), E = 8.400 MPa (N/mm2), density 500 kg/m3

Loading on floor and including floor is 2.5 kPa.

Allowable deflection is L/240.

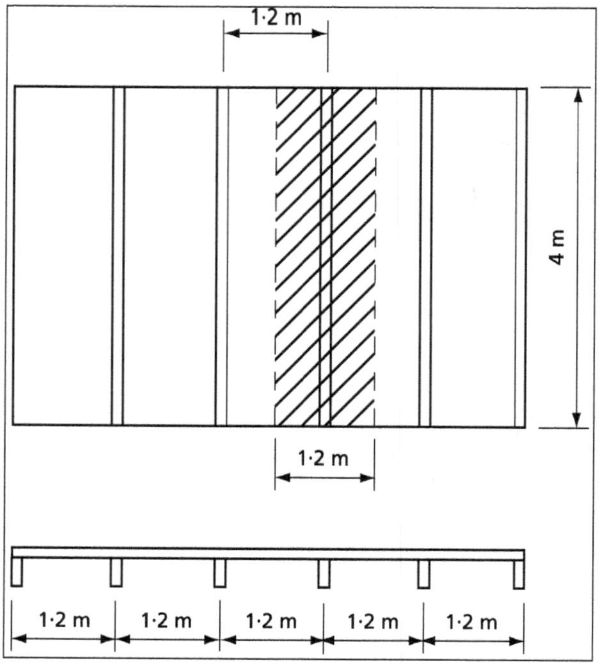

(i) Beam loading: $w = 1.2 \text{ m} \times 2.5 \text{ kN/m}^2 = 3 \text{ kN/m}$.

Assume a 100 mm by 250 mm cross-section for the beams.

(ii) Beam mass = 0.1 × 0.25 × 500 × 9.81 = 122.6 N/m = 0.12 kN/m.

Total $w = 3 + 0.12 = 3.12$ kN/m.

(iii) Calculate reactions and draw shear-force and bending-moment diagrams:

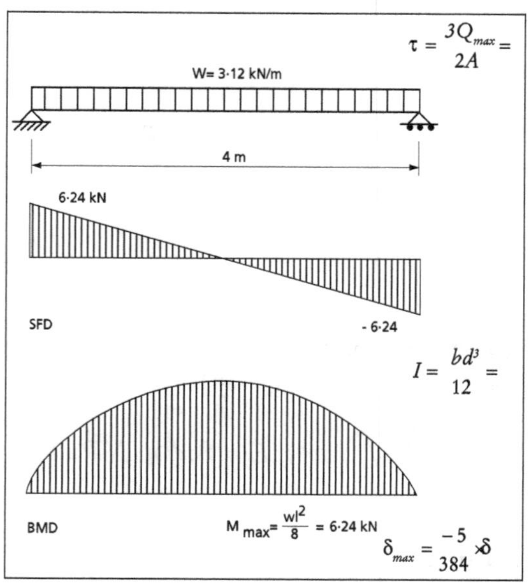

(iv) Calculate maximum bending moment (M_{max}) using the equation for a simple beam, uniformly loaded:

$$M_{max} = \frac{wL^2}{8} = \frac{3.12 \times 4^2}{8} = 6.24 \text{ kNm} = 6.24 \times 10^6 / \text{Nmm}$$

(v) Find the required section modulus (Z):

$$Z_{req} = \frac{M_{max}}{f_w} = \frac{6.24 \times 10^6}{8} = 0.78 \times 10^6 \text{ mm}^3$$

(vi) Find a suitable beam depth, assuming 100 mm breadths:

$$\because d = \sqrt{\frac{6Z}{b}} = \sqrt{\frac{6 \times 0.78 \times 10^6}{100}} = 216 \text{mm}$$

Choose a 100 mm by 225 mm timber. The timber required is a little less than that assumed. No recalculations are required unless it is estimated that a smaller size timber would be adequate if a smaller size had been assumed initially.

(vii) Check for shear loading:

$$\tau = \frac{3Q_{max}}{2A} = \frac{3 \times 6.24 \times 10^3}{2 \times 100 \times 225} = 0.42 \text{MPa}$$

As the safe load for the timber is 0.7 N/mm2 (MPa) the section is adequate in resistance to horizontal shear.

(viii) Check deflection to ensure that it is less than 1 /240 of the span:

$$\delta_{max} = \frac{-5}{384} = \frac{WL^3}{EI}$$

Where,

E = 8400 MPa (N/mm2)

$$I = \frac{bd^3}{12} = \frac{100 \times 225^3}{12} = 95 \times 10^6 \text{ mm}^4$$

W = 3.12 kN/m × 4 m = 12.48 kN = 12.48 × 103 N

L = 4 × 103 mm

$$\delta_{max} = \frac{-5}{384} \times \frac{12.48 \times 10^3 \times 4^3 \times 10^9}{8400 \times 95 \times 10^6} = -13 \text{mm}$$

The allowable deflection, 400/240 = 16.7 >13. The beam is therefore satisfactory.

Bending Moments caused by Askew Loads

If the beam is loaded so that the resulting bending moment is not about one of the main axes, the moment has to be resolved into components acting about the main axes. The stresses are then calculated separately relative to each axis and the total stress is found by adding the stresses caused by the components of the moment.

Example: Design a timber purlin that will span rafters 2.4 m on centre. The angle of the roof slope is 30° and the purlin will support a vertical dead load of 250 N/m and a wind load of 200 N/m acting normal to the roof. The allowable bending stress (f_w) for the timber used is 8 MPa. The timber density is 600 kg/m³.

1. Assume a purlin cross-section size of 50 mm × 125 mm. Find an estimated self-load.

$W = 0.05 \times 0.125 \times 600 \times 9.81 = 37$ N/m

The total dead load becomes 250 + 37 = 287 N/m

2. Find the components of the loads relative to the main axes.

$W_x = 200$ N/m + 287 N/m × cos 30° = 448.5 N/m

$W_y = 287$ N/m × sin 30° = 143.5 N/m

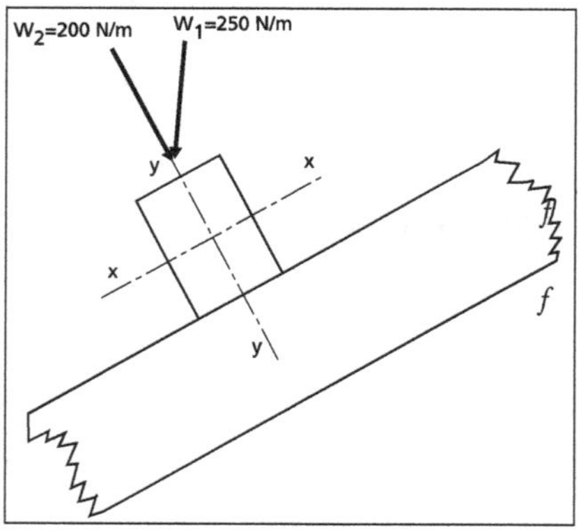

3. Calculate the bending moments about each axis for a uniformly distributed load. The purlin is assumed to be a simple beam.

$$M_{max} = \frac{WL}{8} = \frac{wL^2}{8}$$

$$M_{max\,x} = \frac{w_x \times L^2}{8} = \frac{448.5 \times 2.4^2}{8} = 323 \times 10^3 \text{ Nmm}$$

$$M_{max\,y} = \frac{w_y \times L^2}{8} = \frac{143.5 \times 2.4^2}{8} = 103 \times 10^3 \text{ Nmm}$$

4. The actual stress in the timber must be no greater than the allowable stress.

$$f = \frac{M_{max\,x}}{Z_x} + \frac{M_{max\,z}}{Z_y} \leq f_w$$

5. Try the assumed purlin size of 50 × 125 mm.

$$Z_x = \frac{bd^2}{6} = \frac{50 \times 125^2}{6} = 130 \times 10^3 \text{ mm}^3$$

$$Z_y = \frac{bd^2}{6} = \frac{125 \times 50^2}{6} = 52 \times 10^3 \text{ mm}^3$$

$$f = \frac{323 \times 10^3}{130 \times 10^3} + \frac{103 \times 10^3}{52 \times 10^3} = 2.5 + 2 = 4.5 \text{ N/mm}^2 = 4.5 \text{MPa}$$

This size is safe, but a smaller size may be satisfactory. Try 50 mm × 100 mm.

$$Z_x = \frac{bd^2}{6} = \frac{50 \times 100^2}{6} = 83 \times 10^3 \text{ mm}^3$$

$$Z_y = \frac{bd^2}{6} = \frac{100 \times 50^2}{6} = 42 \times 10^3 \text{ mm}^3$$

$$f = \frac{323 \times 10^3}{83 \times 10^3} + \frac{103 \times 10^3}{42 \times 10^3} = 3.9 + 2.5 = 6.4 \text{ N/mm}^2 = 6.4 \text{MPa}$$

This is much closer to the allowable stress. To save money, 50 mm × 75 mm should also be tried. In this case $f > f_w$ and therefore 50 mm × 100 mm is chosen.

Universal Steel Beams

Sections of steel beams are indicated with a combination of letters and a number, where the letters represent the shape of the section and the number represents the dimension, usually the height, of the section in millimetres, e.g. IPE 100. In the case of HE sections, the number is followed by a letter indicating the thickness of the web and flanges, e.g. HE 180B. An example of an alternative method of notation is 305 × 102 UB 25, i.e. a 305 mm by 102 mm universal beam weighing 25 kg/m.

The following example demonstrates another method of taking into account the self-weight of the structural member being designed.

Example: Design a steel beam, to be used as a lintel over a door opening, which is required to span 4.0 m between centres of simple supports. The beam will be carrying a 220 mm thick and 2.2 m high brick wall, weighing 20 kN/m^3. Allowable bending stress is 165 MPa.

Uniformly distributed load caused by brickwork is 0.22 × 2.2 × 4.0 × 20 = 38.7 kN.

Assumed self-weight for the beam is 1.5 kN. (The triangular load distribution for bricks above the lintel would result in a slightly lower load value). Total uniformly distributed load $W = 38.7 + 1.5 = 40.2$ kN.

$$M_{max} = \frac{WL}{8} = \frac{40.2 \times 4.0}{8} = 20.1 \text{ kNm} = 20.1 \times 10^6 \text{ Nmm}$$

$$Z_{req} = \frac{20.1 \times 10^6}{165} = 0.122 \times 10^6 \text{ mm}^3 = 122 \text{ cm}^2$$

Continuous Beams

A single continuous beam extending over a number of supports will safely carry a greater load than a series of simple beams extending from support to support. Consider the shear force and bending moment diagrams for the following two beam loadings.

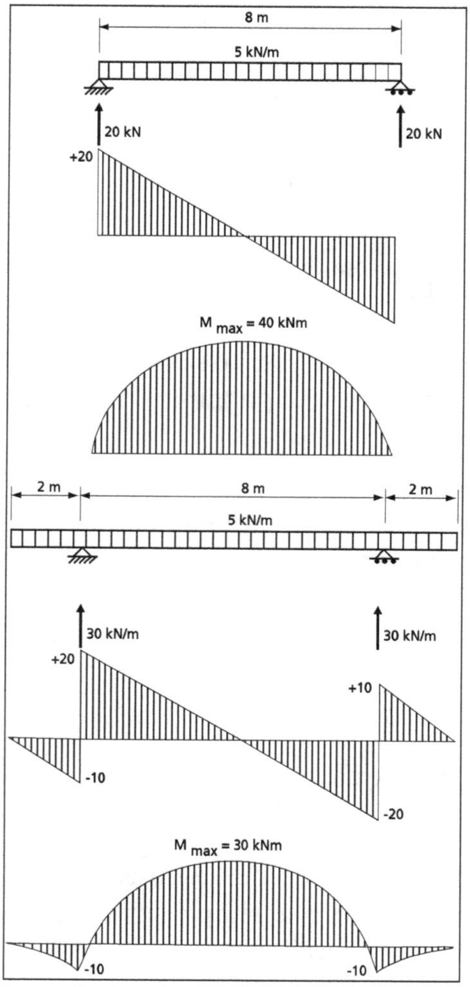

Although the total value of the load has increased, the maximum shear force remains the same but the maximum bending is reduced when the beam is cantilevered over the supports.

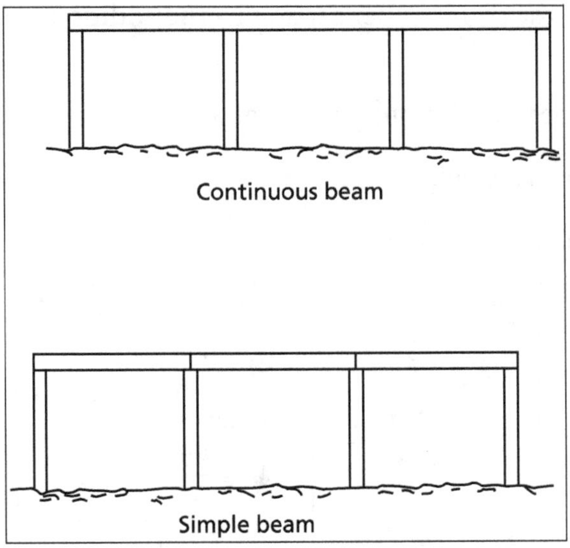

Continuous beam

Simple beam

Although continuous beams are statically indeterminate and the calculations are complex, approximate values can be found with simplified equations. Conservative equations for two situations are as follows:

Load concentrated between supports: $BM = \dfrac{WL}{6}$

Load uniformly distributed: $BM = \dfrac{WL}{12}$

It is best to treat the two end sections as simple beams.

Standard Cases of Beam Loading

A number of beam loading cases occur frequently and it is useful to have standard expressions available for them.

Composite Beams

In small-scale buildings the spans are relatively small and, with normal loading, solid rectangular or square sections are generally the most economical beams. However, where members larger than the available sizes and/or length of solid timber are required, one of the following combinations may be chosen:

- Arranging several pieces of timber or steel into a structural frame or truss.
- Universal steel beams.
- Built-up timber sections with the beam members nailed, glued or bolted together into a solid section, or with the beam members spaced apart and only connected at intervals.

- Strengthening the solid timber section by the addition of steel plates to form a 'flitch-beam'.
- Plywood web beams with one or several webs.
- Reinforced-concrete beams.

Table: Beam Equations.

Loading diagram	Shear force at x: Q_x	Bending moment at x: M_x	Deflection at x: δ_x
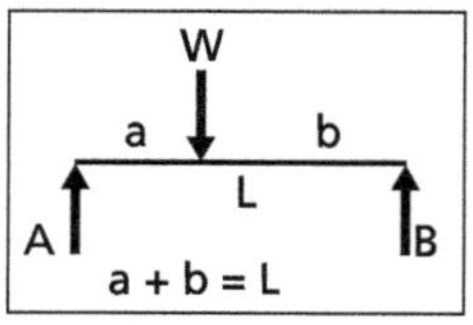	$Q_A = \frac{Wb}{L}$ $Q_B = -\frac{Wa}{L}$	$M_c = \frac{Wab}{L}$ When $a = b$ $M_c = \frac{WL}{4}$	$\delta_c = \frac{Wa^2b^2}{3EIL}$
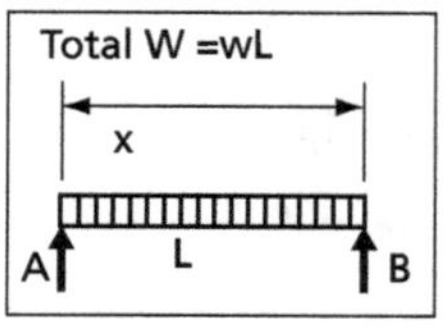	$Q_A = \frac{W}{2}$ $Q_B = -\frac{W}{2}$	$M_{max} = \frac{WL}{8}$ at $x = \frac{L}{2}$	$\delta_{max} = \frac{5WL^3}{384EI}$ at $x = \frac{L}{2}$
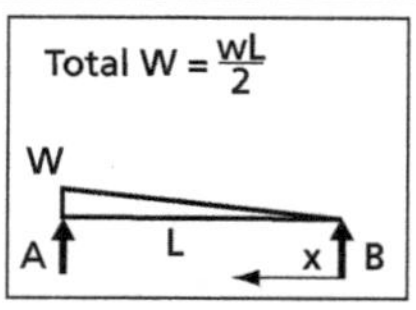	$Q_A = \frac{2W}{3} = \frac{wL}{3}$ $Q_B = -\frac{W}{3} = -\frac{wL}{6}$	$M_{max} = 0.064mL^2$ at $x = 0.577L$	$\delta_{max} = 0.00652\frac{wL^4}{EI}$ at $x = 0.519$
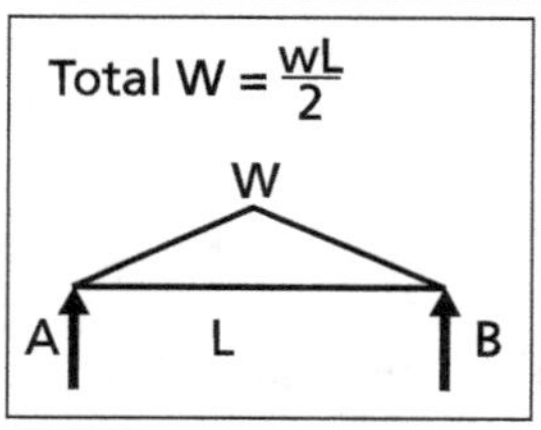	$Q_A = \frac{W}{2} = \frac{wL}{4}$ $Q_B = -\frac{W}{2} = -\frac{wL}{4}$	$M_{max} = \frac{WL^2}{12}$ at $x\frac{L}{2}$	$\delta_{max} = \frac{wL^4}{120EI}$ at $x = \frac{L}{2}$
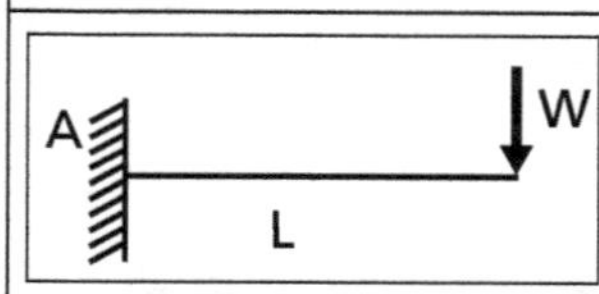	$Q_A = Q_B = W$	$M_A = -WL$	$\delta_B = \frac{WL^3}{3EI}$
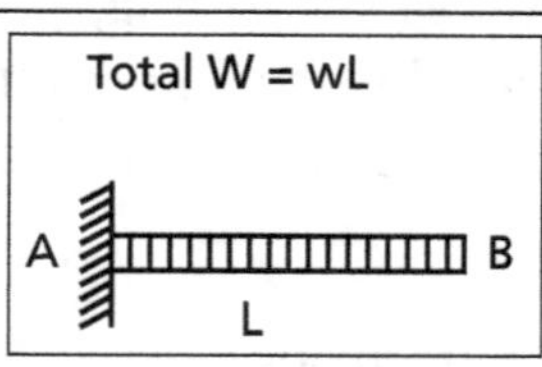	$Q_A = W$ $Q_A = 0$	$M_A = -\frac{WL}{2} = -\frac{wL^2}{2}$	$\delta_B = \frac{WL^3}{3EI} = \frac{wL^4}{3EI}$
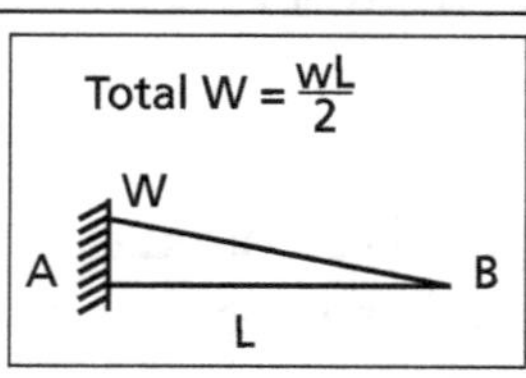	$Q_A = W$ $Q_A = 0$	$M_A = -\frac{WL}{3} = -\frac{wL^2}{6}$	$\delta_B = \frac{wL^4}{30EI}$

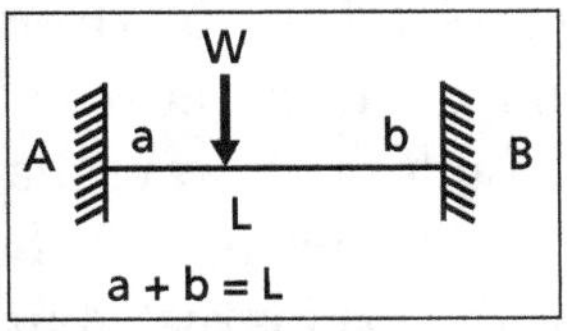	$Q_A = \frac{Wb}{L}$ $Q_B = -\frac{Wa}{L}$	$M_A = -\frac{Wab^2}{L^2}$ $M_B = -\frac{Wa^2b}{L^2}$	$\delta_C = \frac{Wa^3b^3}{3EIL}$
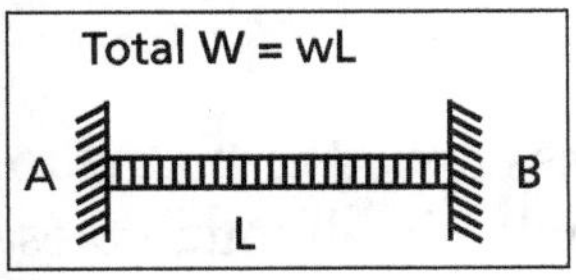	$Q_A = \frac{W}{2}$ $Q_B = -\frac{W}{2}$	$M_A = M_B = -\frac{WL}{12}$	$\delta = \frac{WL}{384EI}$
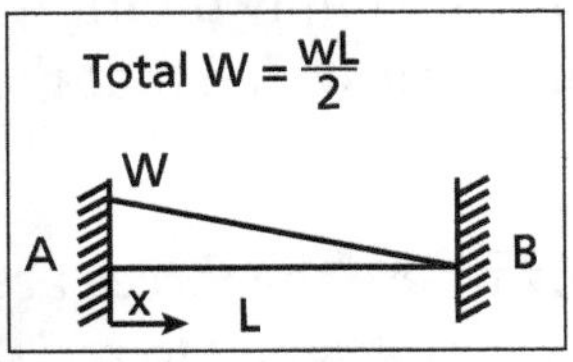	$Q_A = \frac{2W}{3}$ $Q_B = -\frac{W}{3}$	$M_A = -\frac{WL}{10} = -\frac{wL^2}{20}$ $M_B = -\frac{WL}{15} = -\frac{wL^2}{30}$	$\delta_{max} = \frac{wL^4}{764EI}$ at $x = 0.475L$
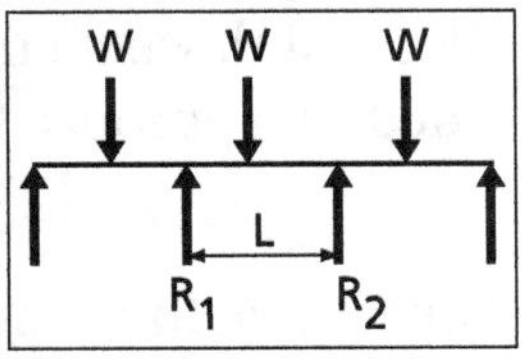	$Q_A = \frac{W}{2}$	$M_{max} = \frac{WL}{6}$ *	$\delta_{max} = \frac{wL^3}{192EI}$
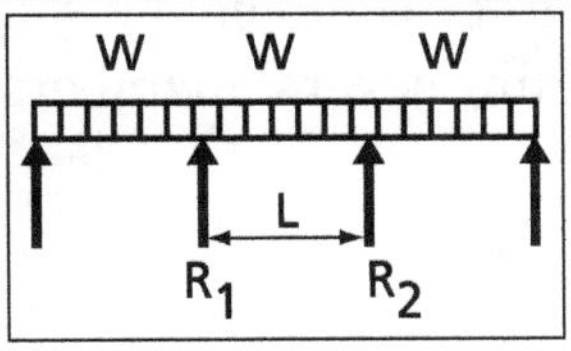	$Q_A = \frac{W}{2}$	$M_{max} = \frac{WL}{12}$ *	$\delta_{max} = \frac{wL^3}{384EI}$

Built-up Timber Beams

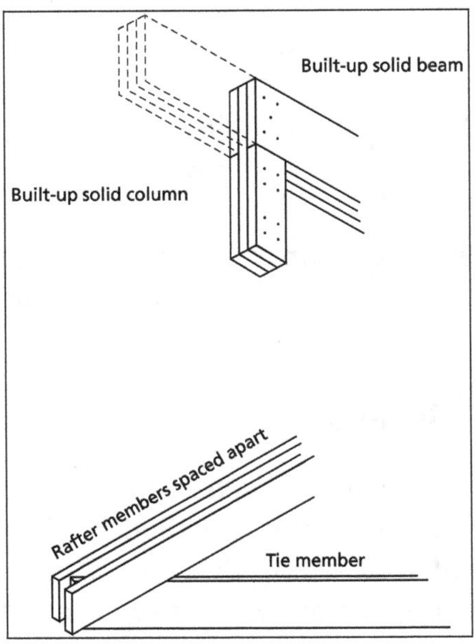

Built-up beams and trusses.

When designing large members, there are advantages in building up solid sections from smaller pieces because these are less expensive and easier to obtain. Smaller pieces also season properly without checking. The composite beams may be built up in ways that minimize warping and permit rigid connections between columns and beams. At the same time the importance of timber defects is decreased, because the load is distributed to several pieces, not all with defects in the same area.

Built-up solid beams are normally formed by using vertical pieces nailed or bolted together: Nailing is satisfactory for beams up to about 250 mm in depth, although these may require the use of bolts at the ends if the shear stresses are high. Simply multiplying the strength of one beam by the number of beams is satisfactory, provided that the staggered joints occur over supports.

Built-up sections with the members spaced apart are used mainly where the forces are tensile, such as in the bottom chords of a truss. Where used in beams designed to resist bending, buckling of the individual members may have to be considered if those members have a large depth-to-width ratio. However, this can be avoided by appropriate spacing of stiffeners that connect the spaced members at intervals.

Where the loading is heavy, the beam will require considerable depth, resulting in a large section modulus to keep the stresses within the allowable limit. If sufficient depth cannot be obtained in one member, it may be achieved by combining several members, such as gluing the members together to form a laminate.

Columns

Although the column is essentially a compression member, the manner in which it tends to fail and the amount of load that causes failure depend on:

- The material of which the column is made.
- The shape of cross-section of the column.
- The end conditions of the column.

The first point is obvious: a steel column can carry a greater load than a timber column of similar size. Columns with a large cross-section area compared with the height are likely to fail by crushing.

Buckling of Slender Columns

If a long, thin, flexible rod is loaded axially in compression, it will deflect a noticeable amount. This phenomenon is called buckling and occurs when the stresses in the rod are still well below those required to cause a compression/shearing-type failure. Buckling is dangerous because it is sudden and, once started, is progressive.

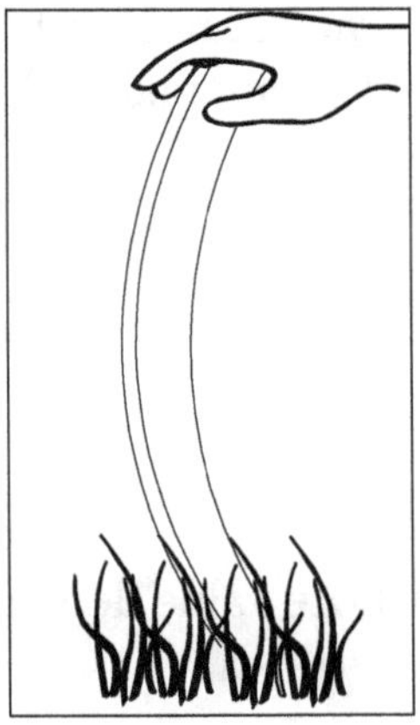

Although the buckling of a column can be compared with the bending of a beam, there is an important difference in that the designer can choose the axis about which a beam bends, but normally the column will take the line of least resistance and buckle in the direction where the column has the least lateral unsupported dimension.

As the loads on columns are never perfectly axial and the columns are not perfectly straight, there will always be small bending moments induced in the column when it is compressed.

There may be parts of the cross-section area where the sum of the compressive stresses caused by the load on the column could reach values larger than the allowable or even the ultimate strength of the material.

Therefore the allowable compressive strength δ_{cw} is reduced by a factor k_λ, which depends on the slenderness ratio and the material used.

$$P_{bw} = K_\lambda \times \delta_{cw} \times A$$

Where,

- P_{bw} = allowable load with respect to buckling,
- k_λ = reduction factor, which depends on the slenderness ratio,
- δ_{cw} = allowable compressive stress,
- A = cross-section area of the column.

When the load on a column is not axial but eccentric, a bending stress is induced in the column as well as a direct compressive stress. This bending stress will need to be considered when designing the column with respect to buckling.

Slenderness Ratio

The relationship between the length of the column, its lateral dimensions and the end fixity conditions will strongly affect the column's resistance to buckling. An expression called the slenderness ratio has been developed to describe this relationship:

$$\lambda = \frac{KL}{r} = \frac{l}{r}$$

Where,

- λ = slenderness ratio,
- K = effective length factor whose value depends on how the ends of the column are fixed,
- L = length of the column,
- r = radius of gyration ($r = I / A$),
- l = effective length of the column ($K \times L$).

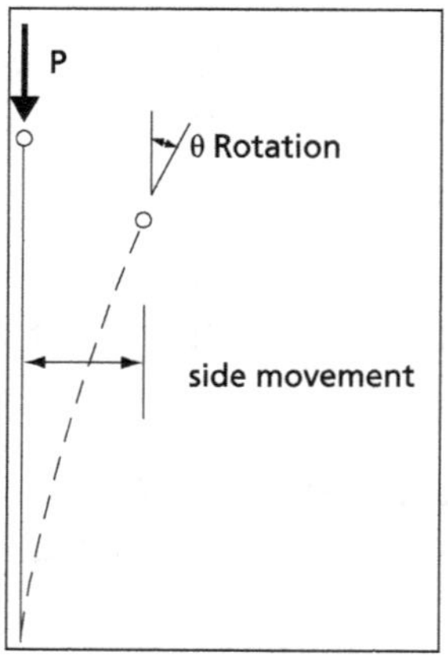

There are four types of end condition for a column or strut:

1. Total freedom of rotation and side movement – like the top of a flagpole. This is the weakest end condition.

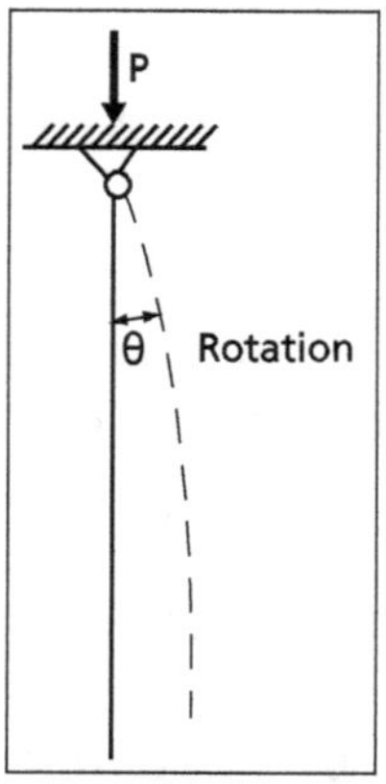

2. Fixed in position but not in direction (pinned).

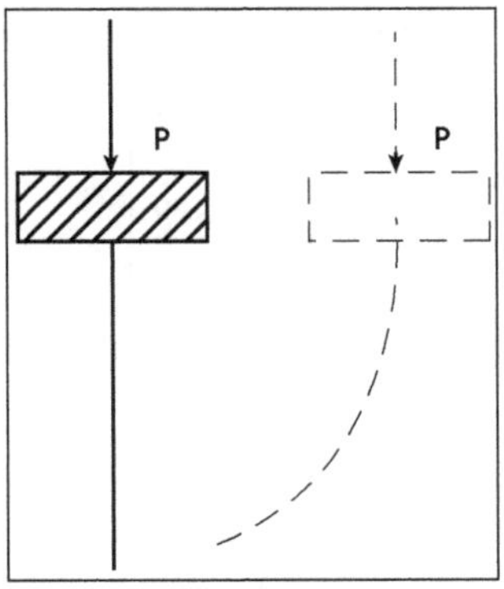

3. Fixed in direction but not in position.

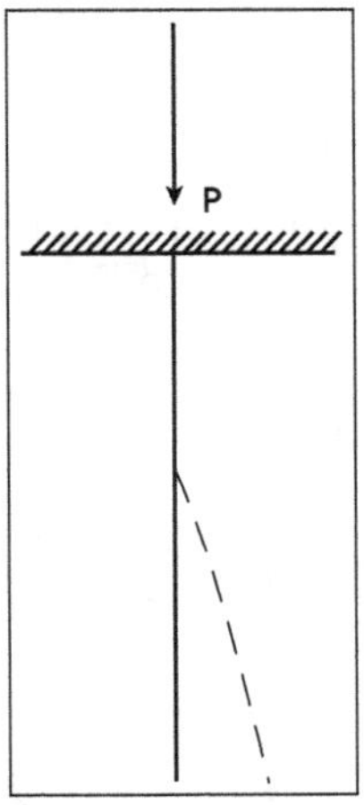

4. Fixed in position and in direction.

The consideration of the two end conditions together results in the following theoretical values for the effective length factor (K_p is the factor usually used in practice).

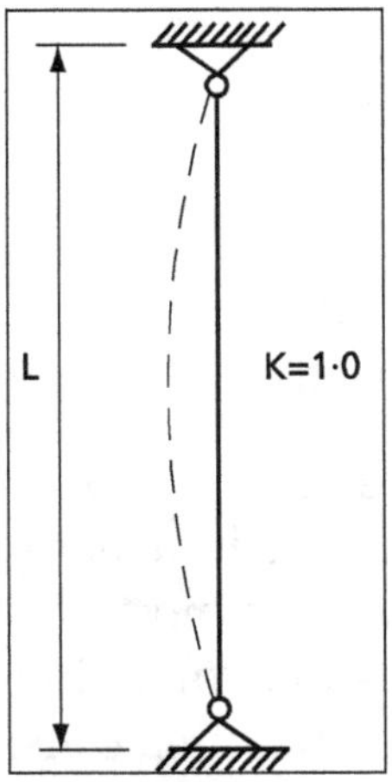

Both ends pinned.

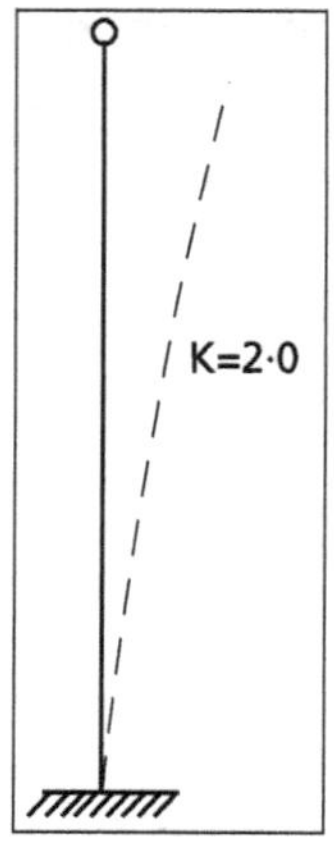

One end fixed in direction and position, the other free.

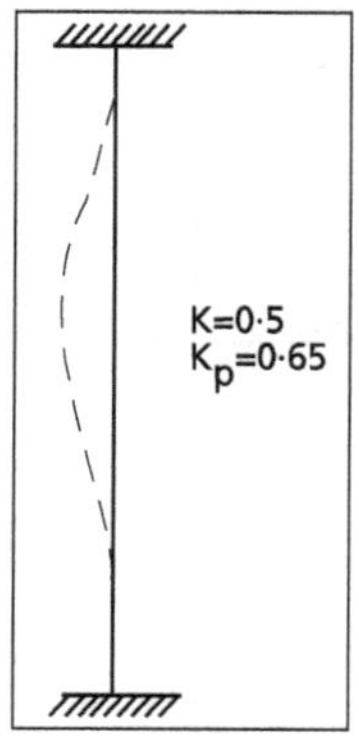

Both ends fixed in direction and position.

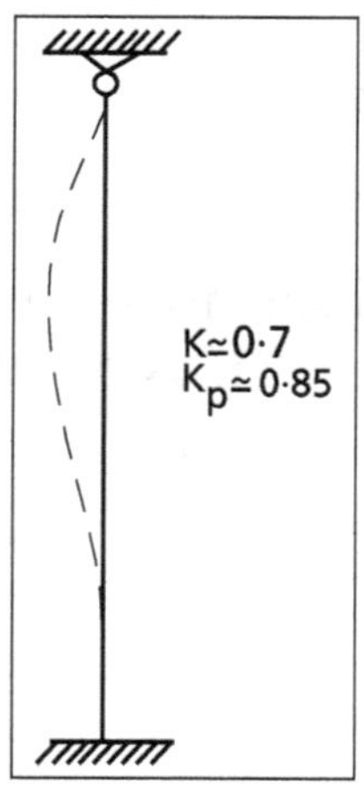

One end pinned, the other fixed in direction and position.

Columns and struts with both ends fixed in position and effectively restrained in direction would theoretically have an effective length of half the actual length. However, in practice this type of end condition is almost never perfect and therefore somewhat higher values for K are used and can be found in building codes. In fact, in order to avoid unpleasant surprises, the ends are often considered to be pinned (K_p = 1.0) even when, in reality, the ends are restrained or partially restrained in direction.

The effective length can differ with respect to the different cross-sectional axes:

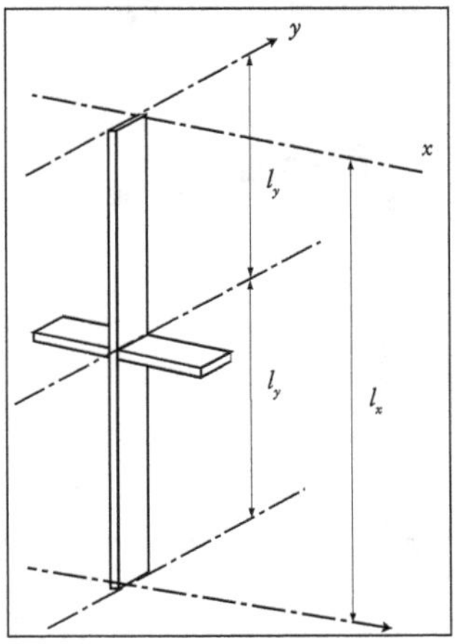

1. A timber strut that is restrained at the centre has only half the effective length when buckling about the y-y axis as when buckling about the x-x axis. Such a strut can therefore have a thickness of less than its width.

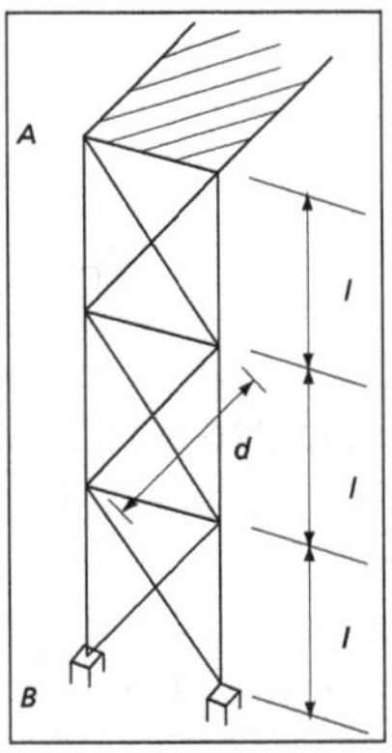

2. In the structural framework, the braces will reduce the effective length to l when the column A-B is buckling sideways but, as there is no bracing restricting buckling forwards and backwards, the effective length for buckling in these directions is $3l$. Similarly, the bracing struts have effective lengths of 1/2 d and d respectively.

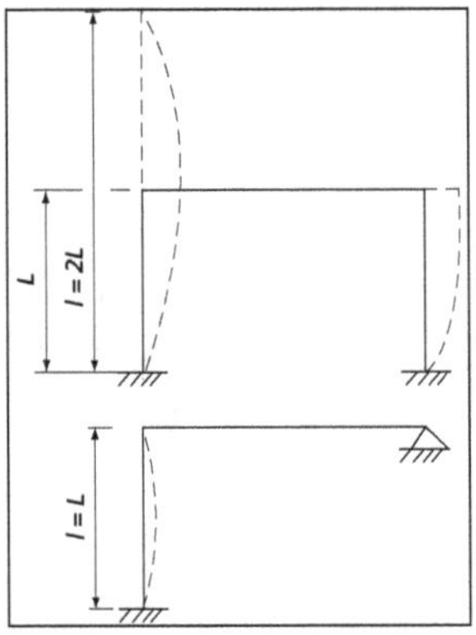

3. The leg of a frame, which is pinned to the foundation, has the effective length l = 2 L but, if the top is effectively secured for sideways movement, the effective length is reduced to $l = L$.

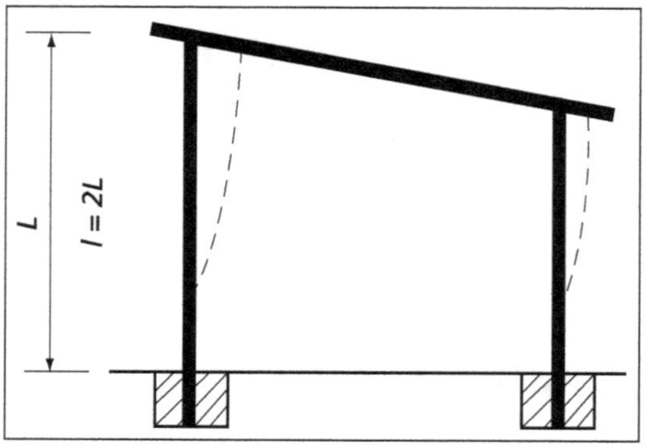

4. In a system of post and lintel where the bottom of the post is effectively held in position and secured in direction by being cast in concrete, the effective length l = 2 L.

Axially Loaded Timber Columns

Timber columns are designed with the following formulae:

$$\lambda = \frac{KL}{r} \text{ and } P_{bw} = k_{\lambda} \times \delta_{cw} \times A$$

In some building codes a value of slenderness ratio in the case of sawn timber is taken as the ratio between the effective length and the least lateral width of the column l/b.

Example: Design a timber column that is 3 metres long, supported as shown in the figure and loaded with a compressive load of 15 kN. Allowable compressive stress (σ_{cw}) for the timber is 5.2 MPa.

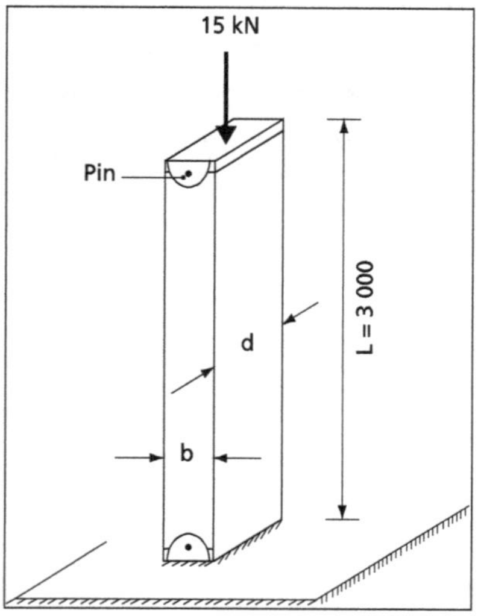

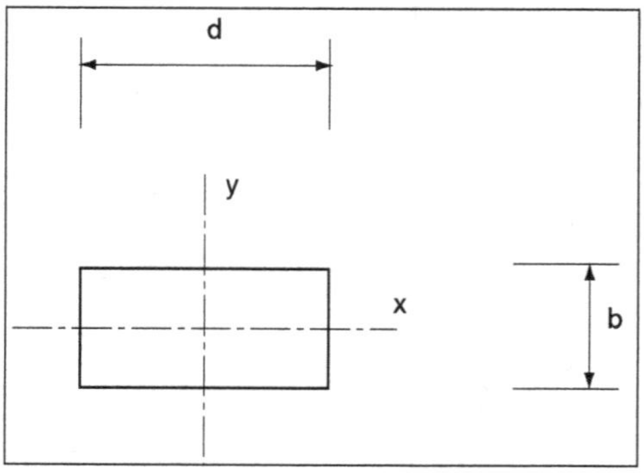

1. In this case, the end conditions for buckling about the *x-x* axis are not the same as about the *y-y* axis. Therefore both directions must be designed for buckling (Where the end conditions are the same, it is sufficient to check for buckling in the direction that has the least radius of gyration).

Find the effective length for buckling about both axes. Buckling about the *x-x* axis, both ends pinned:

$$l_x = 1.0 \times 3000 = 3000\,\text{mm}$$

Buckling about the *y-y* axis, both ends fixed:

$$l_y = 0.65 \times 3000 = 1950\,\text{mm}$$

Table: Reduction factor (k_λ) for stresses with respect to the slenderness ratio for wood columns.

Slender-ness ratio															
	l/b	2.9	5.8	8.7	11.5	14.4	17.3	20.2	23.0	26.0	28.8	34.6	40.6	46.2	52.0
	l/r	10	20	30	40	50	60	70	80	90	100	120	140	160	180
	k_λ	1.0	1.00	0.91	0.81	0.72	0.63	0.53	0.44	0.35	0.28	0.20	0.14	0.11	0.40

2. Choose a trial cross-section, which should have its largest lateral dimension resisting the buckling about the axis with the largest effective length. Try 50 mm × 125 mm. The section properties are:

$$A = b \times d = 50 \times 125 = 6250\,\text{mm}^2$$

$$r_x = \frac{d}{2\sqrt{3}} = \frac{125}{2\sqrt{3}} = 36.1\,\text{mm}$$

$$r_y = \frac{b}{2\sqrt{3}} = \frac{50}{2\sqrt{3}} = 14.4\,\text{mm}$$

3. Find the allowable load with regard to buckling on the column for buckling in both directions.

$$\lambda_x = \frac{l_x}{r_x} = \frac{3000}{36.1} = 83 \text{ gives } k_{\lambda x} = 0.41$$

$$\lambda_y = \frac{l_y}{r_y} = \frac{1950}{14.4} = 135 \text{ gives } k_{\lambda y} = 0.16$$

$$P_w = k_\lambda \times \sigma_c \times A$$

$$P_{wx} = 0.41 \times 5.2 \times 6250\,\text{mm}^2 = 13325\text{ N}$$

$$P_{wy} = 0.16 \times 5.2 \times 6250\,\text{mm}^2 = 5200\text{ N}$$

4. As the allowable load with respect to buckling is smaller than the actual load, a bigger cross-section needs to be chosen. Try 75 mm × 125 mm and repeat steps 2 and 3.

Section properties:

$$A = 75 \times 125 = 9375 \text{ mm}^2$$

$$r_x = \frac{125}{2\sqrt{3}} = 36.1 \text{ mm}$$

$$r_y = \frac{75}{2\sqrt{3}} = 21.7 \text{ mm}$$

Find the allowable buckling load for the new cross-section:

$$\lambda_x = \frac{l_x}{r_x} = \frac{3000}{36.1} = 83 \text{ gives } k_{\lambda x} = 0.41$$

$$\lambda_y = \frac{l_y}{r_y} = \frac{1950}{21.7} = 90 \text{ gives } k_{\lambda y} = 0.16$$

$$P_{wx} = 0.41 \times 5.2 \times 9375 = 19988 \text{ N, say 20 kN}$$

$$P_{wy} = 0.35 \times 5.2 \times 9375 = 17063 \text{ N, say 17 kN}$$

The allowable load with respect to buckling on the column with cross-section 75 mm × 125 mm is therefore 17 kN. Although this is bigger than the actual load, further iterations to find the precise section to carry the 15 kN are not necessary.

The compressive stress in the chosen cross-section will be:

$$\sigma_c = \frac{F}{A} = \frac{15000}{9375} = 1.6 \text{MPa}$$

This is much less than the allowable compressive stress, which makes no allowance for slenderness.

Axially Loaded Steel Columns

The allowable loads for steel columns with respect to buckling can be calculated in the same manner as for timber. However, the relation between the slenderness ratio and the reduction factor (k_λ) is slightly different, as seen in table.

Table: Reduction factor (k_λ) for stresses with respect to the slenderness ratio for steel columns.

λ	10	20	30	40	50	60	70	80	90	100	110	120	130	140
k_λ	0.97	0.95	0.92	0.90	0.86	0.81	0.74	0.67	0.59	0.51	0.45	0.39	0.34	0.30
λ	150	160	170	180	190	200	210	220	230	240	250	300	350	
k_λ	0.26	0.23	0.21	0.19	0.17	0.15	0.14	0.13	0.12	0.11	0.10	0.07	0.05	

Example: Calculate the safe load on a hollow square steel stanchion whose external dimensions are 120 mm × 120 mm. The walls of the column are 6 mm thick and the allowable compressive stress σ_{cw} = 150 MPa. The column is 4 metres high and both ends are held effectively in position, but one end is also restrained in direction.

The effective length of the column l = 0.85L = 0.85 × 4 000 = 3 400 mm.

$$r_x = r_y = \sqrt{\frac{I}{A}} = \sqrt{\frac{BD^3 - bd^3}{12(BD - bd)}} = \sqrt{\frac{120^4 - 108^4}{12(120^2 - 108^2)}} = 46.6$$

$$\lambda = \frac{l}{r} = \frac{3400}{46.6} = 73 \text{ gives } k_\lambda = 0.72 \text{ by interpolation}$$

$$P_w = k_\lambda \times \sigma_{cw} \times A = 0.72 \times 150(120^2 - 108^2) = 295 \text{ kN}$$

Axially Loaded Concrete Columns

Most building codes permit the use of plain concrete only in short columns, that is to say, columns where the ratio of the effective length to least lateral dimension does not exceed 15, i.e. $l/r \leq 15$. If the slenderness ratio is between 10 and 15, the allowable compressive strength must be reduced. The tables of figures relating to l/b in place of a true slenderness ratio are only approximate, as radii of gyration depend on both b and d values in the cross-section and must be used with caution. In the case of a circular column:

$$b = \frac{D}{4} \times \sqrt{12} \approx 0.87D, \text{ where}$$

D = the diameter of the column.

Table: Permissible compressive stress (P_{cc}) in concrete for columns (MPa or N/mm²).

Concrete mix	Slenderness ratio, l/b					
	≤ 10	11	12	13	14	15
Prescribed						
C10	3.2	3.1	3.0	2.9	2.8	2.7
C15	3.9	3.8	3.7	3.6	3.5	3.4
C20	4.8	4.6	4.5	4.3	4.2	4.1
Nominal						
1:3:5	3.1	3.0	2.9	2.8	2.7	2.6
1:2:4	3.8	3.7	3.6	3.5	3.4	3.3
1:1.5:3	4.7	4.5	4.4	4.2	4.1	4.0

Example: A concrete column with an effective length of 4 metres has a cross-section of 300 mm by 400 mm. Calculate the allowable axial load if a nominal 1:2:4 concrete mix is to be used.

Slenderness ratio:

$$\frac{l}{b}=\frac{4000}{300}=13.3$$

Hence, table gives P_{cc} = 3.47 N/mm² by interpolation.

$$P_w = P_{cc} \times A = 3.47 \times 300 \times 400 = 416.4 \text{ kN}$$

Eccentrically Loaded Timber and Steel Columns

Where a column is eccentrically loaded, two load effects need to be considered: The axial compressive stress caused by the load and the bending stresses caused by the eccentricity of the load.

Obviously, by the law of superposition, the added stresses of the two load effects must be below the allowable stress.

Therefore, $\frac{\sigma}{P_{cw}}+\frac{f}{f_w}\leq 1$ i.e.

$$\frac{\text{axial compressive stress}}{\text{allowable compressive stress}}+\frac{\text{bending stress}}{\text{allowable bending stress}}\leq 1$$

$\frac{\sigma_c}{k_\lambda \times \sigma_{cw}}+\frac{f}{f_w}\leq 1$ which can be transferred to:

$$\frac{P_1}{K_\lambda \times A}+\frac{\sigma_{cw}}{f_w}\times\frac{M}{Z}\leq \sigma_{cw}$$

Example: Determine within 25 mm the required diameter of a timber post loaded as shown in the figure. The bottom of the post is fixed in both position and direction by being cast in a concrete foundation. Allowable stresses for the timber used are σ_{cw} = 9 MPa and f_w = 10 MPa.

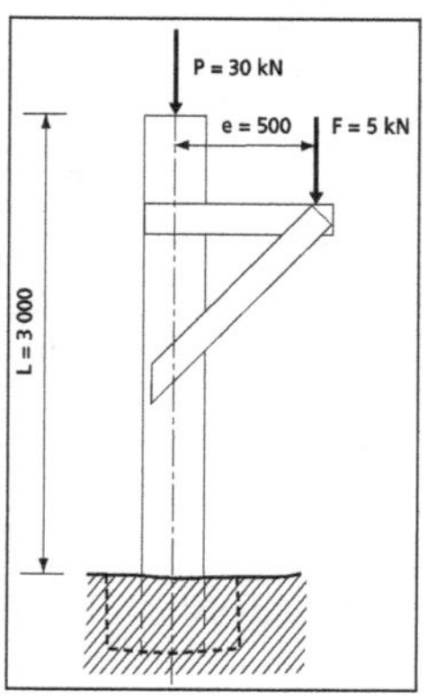

The load of 5 kN on the cantilever causes a bending moment of $M = F \times e$ = 5 kN × 0.5 m = 2.5 kNm.

The effective length of the post = $L \times K$ = 3 000 × 2.1 = 6 300 mm. Try a post with a diameter of 200 mm.

The cross-sectional properties are:

$$A = \frac{\pi D^2}{4} + \frac{\pi \times 200^2}{4} = 31400 \text{ mm}^2$$

$$Z = \frac{\pi D^3}{32} + \frac{\pi \times 200^3}{32} = 785400 \text{ mm}^2$$

$$r = \frac{D}{4} + \frac{200}{4} = 50 \text{ mm}$$

The slenderness ratio = $\frac{l}{r} + \frac{6300}{50} = 126$

Interpolation in table gives k_λ = 0.18

$$\frac{P_1}{K_\lambda \times A} + \frac{\sigma_{cw}}{f_w} \times \frac{M}{Z} \leq \sigma_{cw}$$

$$\frac{30000}{0.18 \times 31400} + \frac{9}{10} \times \frac{2.5 \times 10^6}{785400} = 8.17 \text{ N/mm}^2 \leq 9 \text{ N/mm}^2$$

If the post has a diameter of 200 mm, it will be able to carry the loads, but the task was to determine the diameter within 25 mm. Therefore a diameter of 175 mm should also be tried.

$$\lambda = \frac{6300}{43.75} = 144$$

0.13

$$\frac{30000}{0.13 \times 24050} + \frac{9}{10} \times \frac{2.5 \times 10}{167480} = 23 \text{ N/mm}^2 > 9 \text{ N/mm}^2$$

This diameter is too small, so a diameter of 200 mm should be chosen. It will be appreciated that the choice of effective length based on end fixity has a great effect on the solution.

Plain and Centrally Reinforced Concrete Walls

Basically walls are designed in the same manner as columns, but there are a few differences. A wall is distinguished from a column by having a length that is more than five times the thickness.

Plain concrete walls should have a minimum thickness of 100 mm. Where the load on the wall is eccentric, the wall must have centrally placed reinforcement of at least 0.2 percent of the cross-section area if the eccentricity ratio *e/b* exceeds 0.20. This reinforcement may not be included in the load-carrying capacity of the wall.

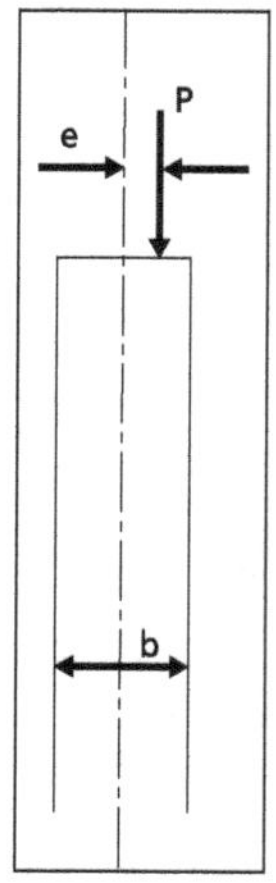

Many agricultural buildings have walls built of blocks or bricks. The same design approach as that shown for plain concrete with axial loading can be used. The maximum allowable compressive stresses must be ascertained, but the reduction ratios can be used as before.

Example: Determine the maximum allowable load per metre of a 120 mm thick wall, with an effective height of 2.8 metres and made from concrete grade C 15: (a) when the load is central; (b) when the load is eccentric by 20 mm:

Slenderness ratio, $\frac{l}{b}=\frac{2800}{120}=23.3$

Interpolation gives:

$$P_{cw}=2.8-\frac{3.3}{5}(2.8-2.0)=2.27\ \text{N/mm}^2=2.27\ \text{MPa}$$

Allowable load:

$$\begin{aligned}P_w&=A\times P_{cw}=1.0\times 0.12\times 2.27\times 10^6\\&=272.4\ \text{kN/m wall}\end{aligned}$$

Ratio of eccentricity $\frac{e}{b}=\frac{20}{120}=0.167$

A double interpolation gives:

$$P_{cw}=1.06\ \text{N/mm}^2=1.06\ \text{MPa}$$

Allowable load:

$$P_w = A \times P_{cw} = 1.0 \times 0.12 \times \frac{1.06 \times 10^6}{1000} = 127.2 \text{ kN/m wall}$$

Central reinforcement is not required because $\frac{e}{b} < 20$.

Table: Allowable compressive stress, P_{cw} for concrete used in walls (N/mm2).

Concrete grade or mix	Slenderness ratio l/b	Ratio of eccentricity of the load e/b					
		Plain concrete walls			Centrally reinforced concrete walls		
		0.00	0.10	0.20	0.30	0.40	0.50
C20	25	2.4	1.7	0.9	-	-	-
	20	3.3	2.3	1.4	0.8	0.4	0.3
	15	4.1	3.0	2.0	0.8	0.4	0.3
	≤ 10	4.8	3.7	2.7	0.8	0.4	0.3
C15	25	2.0	1.3	0.7	-	-	-
	20	2.8	1.9	1.1	0.7	0.7	0.25
	15	3.4	2.4	1.7	0.7	0.7	0.25
	≤ 10	3.9	3.0	2.2	0.7	0.7	0.25
C10	20	2.3	1.6	1.0	0.5	0.3	0.2
	15	2.7	2.0	1.4	0.5	0.3	0.2
	≤ 10	3.2	2.5	1.8	0.5	0.3	0.2
1:1:3	25	2.3	1.6	0.8	-	-	-
1:1:2	20	3.2	2.2	1.3	0.8	0.4	0.3
	15	4.1	2.9	1.9	0.8	0.4	0.3
	≤ 10	4.7	3.6	2.6	0.8	0.4	0.3
1:2:3	20	3.0	2.1	1.3	0.7	0.35	0.25
	15	3.7	2.7	1.9	0.7	0.35	0.25
	≤ 10	4.3	3.4	2.5	0.7	0.35	0.25
1:2:4	20	2.7	1.8	1.0	0.6	0.3	0.2
	15	3.3	2.3	1.6	0.6	0.3	0.2
	≤ 10	3.8	2.9	2.1	0.6	0.3	0.2

Trusses

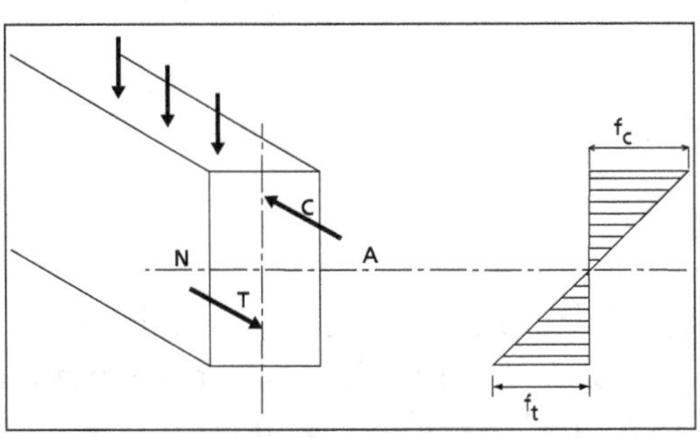

It can be seen from the stress distribution of a loaded beam that the greatest stress occurs at the top and bottom extremities of the beam.

This led to the improvement on a rectangular section by introducing the I-section in which the large flanges were situated at a distance from the neutral axis. In effect, the flanges carried the bending in the form of tension stress in one flange and compression stress in the other, while the shear was carried by the web.

For these situations where bending is high but shear is low, for example in roof design, material can be saved by raising a framework design. A truss is a pinpointed framework.

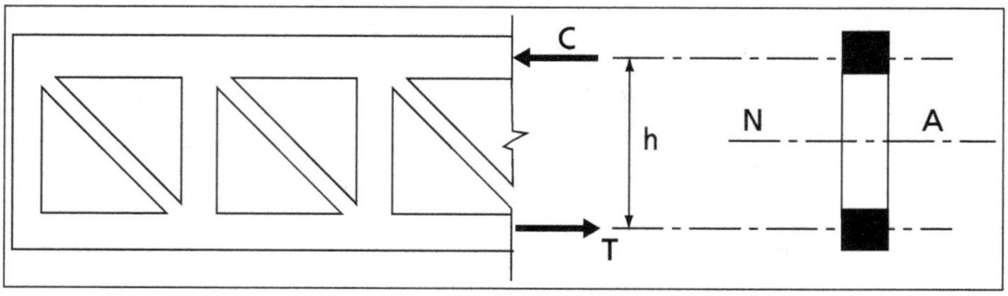

A truss concentrates the maximum amount of materials as far away as possible from the neutral axis. With the resulting greater moment arm (*h*), much larger moments can be resisted.

Resistance of a truss at a section is provided by:

$$M = C \times h = T \times h$$

Where,

- C = T in parallel chords.
- C = compression in the top chord of the truss.
- T = tension in bottom chord of a simply supported truss.
- h = vertical height of truss section.

If either *C* or *T* or h can be increased, then the truss will be capable of resisting heavier loads. The value of h can be increased by making a deeper truss.

Allowable *C*- or *T*-stresses can be increased by choosing a larger cross-section for the chords of the truss, or by changing to a stronger material.

A framework or truss can be considered as a beam with the major part of the web removed. This is possible where bending stresses are more significant than shear stresses. The simple beam has a constant section along its length, yet the bending and shear stresses vary. The truss, comprising a number of simple members, can be fabricated to take into account this change in stress along its length.

The pitched-roof truss is the best example of this, although the original shape was probably designed to shed rainwater. Roof trusses consist of sloping rafters that meet at the ridge, a main tie connecting the feet of the rafters and internal bracing members. They are used to support a roof covering in conjunction with purlins, which are laid longitudinally across the rafters, with the roof cover attached to the purlin. The arrangement of the internal bracing depends on the span.

Rafters are normally divided into equal lengths and, ideally, the purlins are supported at the joints so that the rafters are only subjected to axial forces. This is not always practicable because purlin spacing is dependent on the type of roof covering. When the purlins are not supported at the panel joints, the rafter members must be designed for bending as well as axial force.

The internal bracing members of a truss should be triangulated and, as far as possible, arranged so that long members are in tension and compression members are short to avoid buckling problems.

The outlines in figure give typical forms for various spans. The thick lines indicate struts. The lattice girder, also called a truss, is a plane frame of open web construction, usually with parallel chords or booms at top and bottom. There are two main types, the N- (or Pratt) girder and the Warren girder. They are very useful in long-span construction, in which their small depth-to-span ratio, generally about 1 /10 to 1 /14, gives them a distinct advantage over roof trusses. Steel and timber trusses are usually designed assuming pin-jointed members. In practice, timber trusses are assembled with bolts, nails or special connectors, and steel trusses are bolted, riveted or welded. Although these rigid joints impose secondary stresses, it is seldom necessary to consider them in the design procedure. The following steps should be considered when designing a truss:

- Select general layout of truss members and truss spacing.
- Estimate external loads to be applied including self-weight of truss, purlins and roof covering, together with wind loads.
- Determine critical (worst combinations) loading. It is usual to consider dead loads alone, and then dead and imposed loads combined.
- Analyse the framework to find forces in all members.
- Select the material and section to produce in each member a stress value that does not exceed the permissible value. Particular care must be taken with compression members (struts), or member's normally in tension but subject to stress reversal caused by wind uplift.

Unless there are particular constructional requirements, roof trusses should, as far as possible, be spaced to achieve minimum weight and economy of materials used in the total roof structure. As the distance between trusses is increased, the weight of the purlins tends to increase more rapidly than that of the trusses. For spans up to around 20 m, the spacing of steel trusses is likely to be about 4 metres and, in the case of timber, 2 metres.

The pitch, or slope, of a roof depends on locality, imposed loading and type of covering. Heavy rainfall may require steep slopes for rapid drainage; a slope of 22° is common for corrugated steel and asbestos roofing sheets. Manufacturers of roofing material usually make recommendations regarding suitable slopes and fixings.

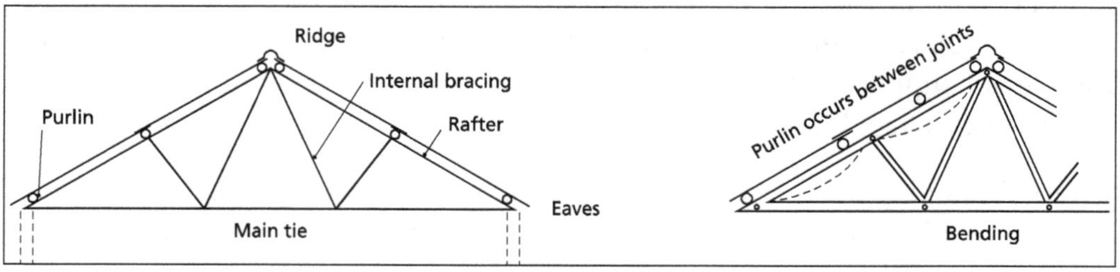

Truss components.

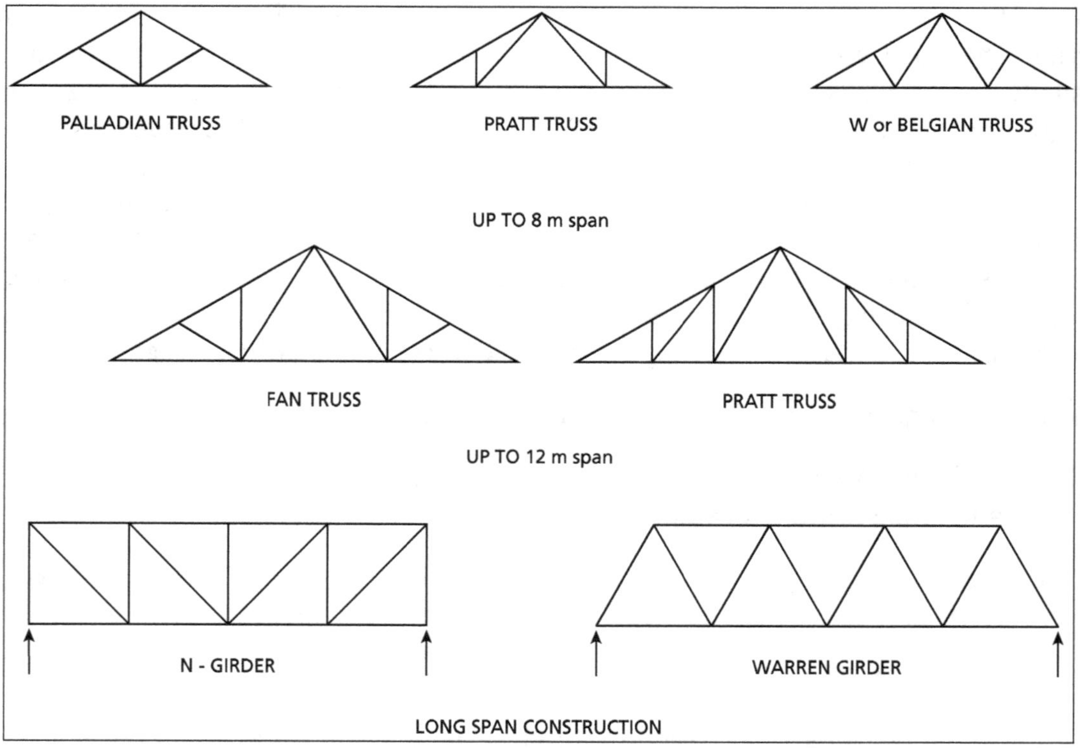

Types of trusses.

A simplified approach can be taken if the intention is to use a common section throughout. Once the layout has been chosen, the member that will carry the maximum load can be established. An understanding of the problems of instability of compression members will lead the designer to concentrate on the top chord or rafter members. A force diagram or method of sections can then be used to determine the load on these members and the necessary size.

Example: A farm building comprising block walls carries steel roof trusses over a span of 8 metres. Roofing sheets determine the purlin spacing. Design the roof trusses.

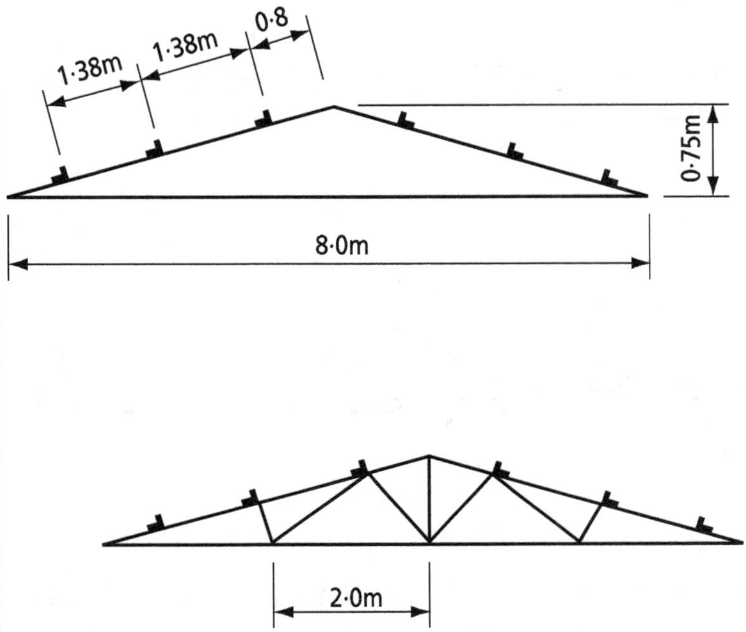

LAYOUT CHOSEN
(nodes at purlin points)

Assume a force analysis shows maximum rafter forces of approximately 50 kN in compression (*D* + *I*) and 30 kN in tension (*D* + *W*), outer main tie member 50 kN tension (*D* + *I*) and 30 kN compression (*D* + *W*). A reversal of forces caused by the uplift action of wind will cause the outer main tie member to have 50 kN of tension and 30 kN of compression.

- Rafter: Using two angles back-to-back will be satisfactory because the distance between restraints is only 1.38 m. (Note that angles must be battened together along the length of the rafter).

- Main Tie: The 65 mm × 50 mm × 6 mm section can carry the required tensile force. Although its length is a little greater than 1.8 m, the compressive load brought about by wind uplift is safe as the design codes allow a greater slenderness ratio for intermittent loads such as wind.

- Finished Design: Note the use of a sole plate to safely distribute the load to the blockwork wall to ensure that the bearing stress of the blocks is not exceeded.

Frames

Apart from the roof truss, there are a number of other structural frames commonly used in farm building construction. They include portal frames, pole barns and post-and-beam frames.

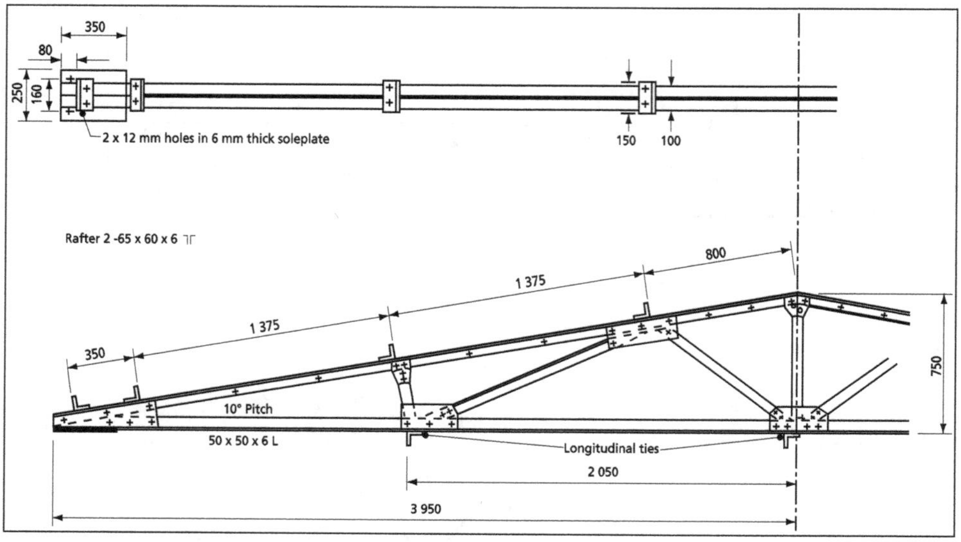

Finished design of the roof truss.

In figure, all welds to be 4mm fillet. All bolts to be M16. Gusset plates to be 8mm thick. Internal bracing shown 65 x 50 x 6 to use common section (size can be reduced if others available). All sections in grade 43 steel. Purlin supports: 70 x 70 x 6 with 2 x 6 ø holes.

A single-bay portal frame consists of a horizontal beam or pitched rafters joined rigidly to vertical stanchions on either side to form a continuous plane frame. For design purposes, portal frames can be classified into three types: fixed base, pinned base (two pins), pinned base and ridge (three pins).

The rigid joints and fixed bases have to withstand bending moments and all bases are subjected to horizontal, as well as vertical, reactions. Hence foundation design requires special attention. The externally applied loads cause bending moments, shear forces and axial forces in the frame.

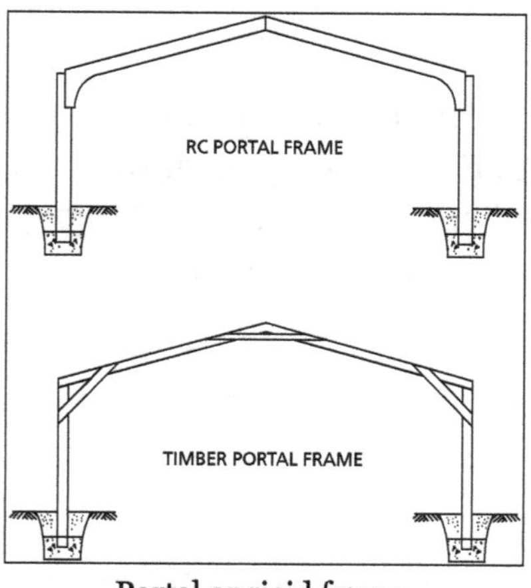

Portal or rigid frame.

Portal frames may be made of steel, reinforced concrete or timber. With wider spans the structural components become massive if timber or reinforced concrete is used. Hence, steel frames are most common for spans over 20 m. At the eaves, where maximum bending moments occur, the section used will need a greater depth than at other points in the frame.

Pole barns are usually built with a relatively simple foundation, deeper than usual, and backfilled with rammed earth. Pole barns are braced between columns and rafters in each direction. The braces serve to reduce the effective length of compression members and the effective span of rafters and other beam members. This leads to a structure that is simple to analyse and design, and can be a low-cost form of construction.

A shed-type building is a simple construction consisting of beams (horizontal or sloping), supported at their ends on walls or posts. There may be one or more intermediate supports depending on the width of the building. Purlins running longitudinally support the roof cover.

As the principal members are simple or continuous beams (very often timber of rectangular section), the stress-analysis aspect of the design is straightforward. When the beam is supported by timber posts, the design of the posts is not difficult because the load is assumed to be axial. Like the poles in the pole barn, the foundation can consist of a simple pad of concrete beneath the post, or the base of the post can be set into concrete.

Example: Design the roof of a building using block walls, timber posts and rafters as shown in the figures.

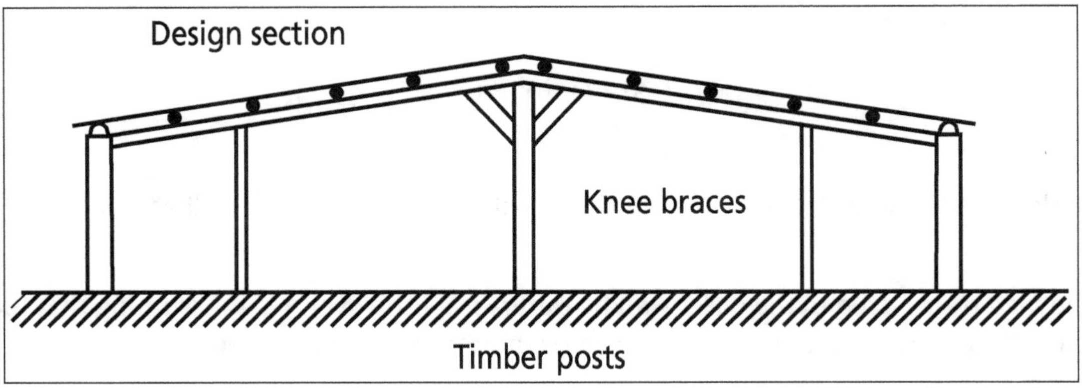

It is assumed that the knee braces reduce the effective span of the rafters between the central wall and the timber posts.

The moments and forces involved are as shown in the diagram. Self-weights and service load have been estimated. Continuity over post and brace has been disregarded. This provides a simple but safe member.

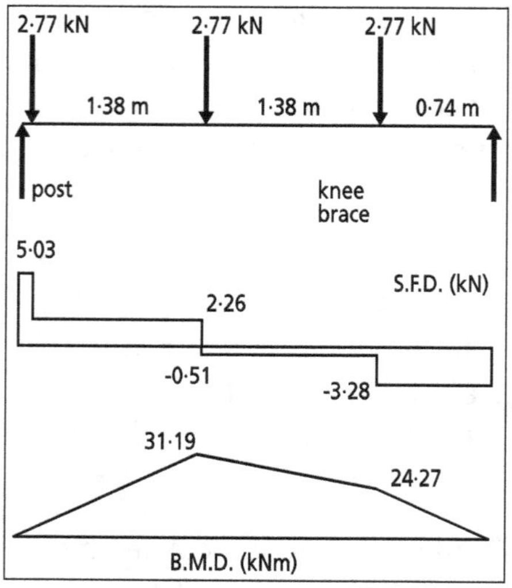

Self-weights and service load have been estimated. Continuity over post and brace has been disregarded. This provides a simple safe member.

Maximum shear force = 5 kN

Maximum bending moment = 3 120 kN/ mm2.

Try two rafters at 38 × 200 (back to back):

$$\text{Maximum shear stress} = \frac{3Q}{2bd} = \frac{3}{2} \times \frac{5000}{76 \times 200}$$

$$= 0.49\ \text{N/mm}^2 = 0.49\ \text{MPa}$$

$$\text{Maximum bending stress} = \frac{M_y}{I} = \frac{M}{Z}$$

$$= \frac{3120 \times 10^3 \times 6}{76 \times 200^2} = 6.2\ \text{N/mm}^2$$

Tables of allowable stresses indicate that most hardwoods, but not all softwoods, are adequate.

The load transferred to the outer wall by rafters is a little over 3 kN. Assuming that the strength of the blocks is at least 2.8 MPa (N/mm2), the area required is:

$$\frac{3000}{2.8} = 1072\,\text{mm}^2$$

As the rafter underside is 76 mm, the minimum interface across the wall is:

$$\frac{1072}{76} = 14\,\text{mm}$$

Hence there is no problem of load transfer to the wall.

Assume posts are 100 mm × 100 mm and 2.5 m long, then $l / b = 25$ and table gives $K = 0.38$.

With $\sigma_c = 5.2$ MPa (N/mm^2) allowable for design, $0.38 \times 5.2 \text{ N/mm}^2 \times 100^2 \approx 20$ kN.

Therefore the load is within the safety margin.

Connections

Timber Structure

The methods used to join members include lapped and butt connectors. Bolt and connector joints, nailed joints and glued joints, and sometimes a combination of two types, are examples of lapped connections. Butt connections require the use of plates or gussets. In all cases the joints should be designed by calculating the shear forces that will occur in the members.

If two members overlap, the joint is called a single lap joint. If one is lapped by two other members, i.e. sandwiched between them, it is called a double-lap joint.

With a single lap, the joint is under eccentric loading. While for small-span trusses carrying light loads this is not significant, when the joints carry large loads eccentricity should be avoided by the use of double-lap joints. Double members are also used to obtain a satisfactory arrangement of members in the truss as a whole.

Sandwich construction enables the necessary sectional area of a member to be obtained by the use of relatively thin timbers, with any double members in compression being blocked apart and fixed in position to provide the necessary stiffness.

Butt Joints

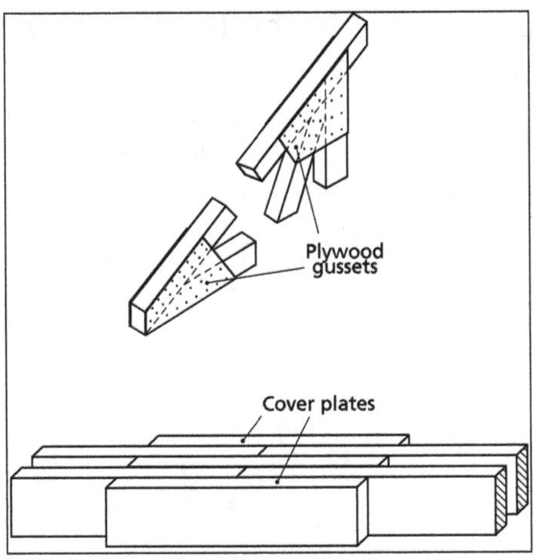

Butt joints.

The use of gussets permits members to butt against each other in the same plane avoids eccentric loading on the joints and, where necessary, provides a greater joining area than is possible with lapped members. This is often an important factor in nailed and glued joints. Arrangement of members on a single centre line is usually possible with gussets.

When full-length timber is not available for a member, a butt joint with cover plates can be used to join two pieces together. This should be avoided, if possible, for the top members (rafters) of a truss and positioned near mid-span for the bottom member (main tie).

Bolt and Connector Joints

Simple bolted joints should only be used for lightly loaded joints because the bearing area at the hole (hole diameter × member thickness) and the relatively low bearing stress allowed for the timber compared with that of the steel bolt, may cause the timber hole to elongate and fail.

Timber connectors are metal rings or toothed plates used to increase the efficiency of bolted joints. They are embedded half into each of the adjacent members and transmit loads from one to the other. The type most commonly used for light structures is the toothed-plate connector, a mild-steel plate cut and stamped to form triangular teeth projecting on each side that embed in the surfaces of the members on tightening the bolt that passes through the joint. The double-sided toothed connector transmits the load and the bolt is assumed to take no load.

Glued Joints

Glues made from synthetic resins produce the most efficient form of joint, as strong as or even stronger than the timber joint, and many are immune to attack by dampness and decay. With this type of joint, all contact surfaces must be planed smooth and the necessary pressure applied during setting of the glue. Bolts or nails that act as clamps are often used and left in place.

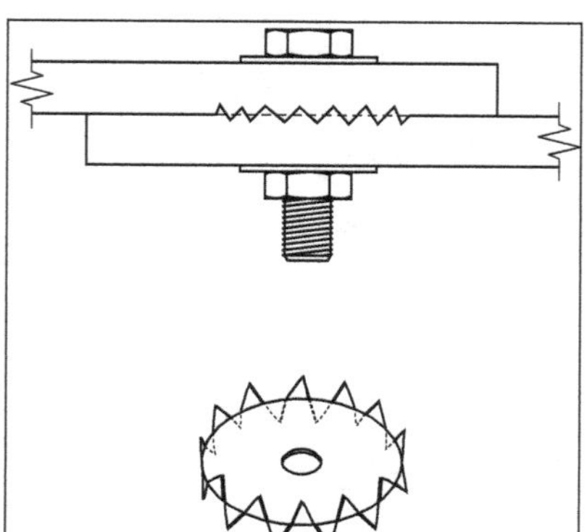

Double-sided toothed plate connector.

The members may be glued directly to each other using lapped joints, or single-thickness construction may be used by the adoption of gussets. As with nailed joints, lapped members may not provide sufficient gluing area and gussets must then be used to provide the extra area.

Glued joints are more often used when trusses are prefabricated because control over temperature, joint fit and clamping pressure is essential. For home use, glue is often used together with nail joints.

Nailed Joints

Although joining by nails is the least efficient of the three methods mentioned, it is an inexpensive and simple method, and can be improved upon by using glue in combination with the nails.

When trusses are prefabricated in factories, nailing plates are often used to connect the member. These fasteners come in two types:

- A thin-gauge plate called a pierced-plate fastener, which has holes punched regularly over its surface to receive nails. The pierced plate can also be used for on-site fabrication.
- A heavier plate with teeth punched from the plate and bent up 90 degrees, called a toothed-plate fastener, or connector. This type, in which the teeth are an integral part of the plate, must be driven in by a hydraulic press or roller.

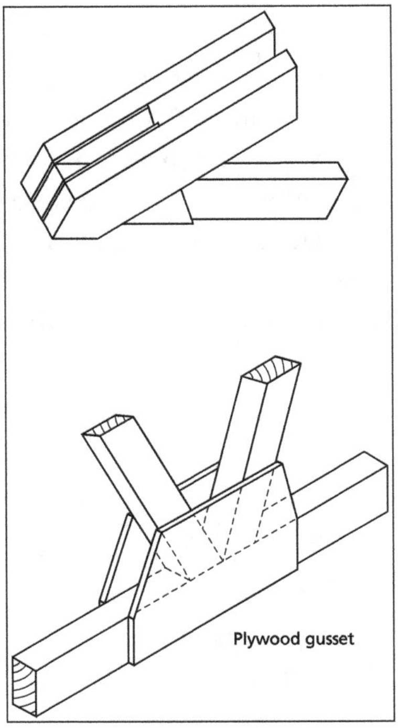

Truss gussets.

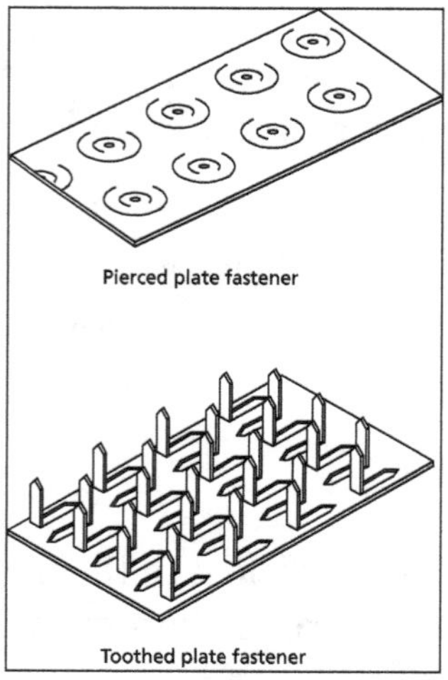

Nailing plates for truss construction.

Table: Minimum nailing distances.

Nailing Area						
x	r_o	d_1	d_{11}	r_b	e_o	e_b
0	5d	5d	10d	5d	-	15d
10	5d	5d	10d	5.5d	8d	15d
20	5d	5d	10d	6d	8d	15d
30	5d	5d	10d	6.5d	8d	15d
40	5d	5d	10d	7d	8d	15d
50	5d	5d	10d	7.5d	8d	15d
≤ 60	5d	5d	10d	8d	8d	15d

d: Diameter of the nail (mm). r_0: Distance from the extreme row of nails to the unloaded edge of member. d_1: Distance between two nails in the nailing area, measured perpendicular to the axis of the member. d_{11}: Distance between two nails measured parallel to the axis of the member. r_b: Distance from the extreme row of nails to the loaded edge of the member. e_0: Distance from the nearest row of nails to the unloaded end of member. e_b: Distance from the nearest row of nails to the loaded end of the member.

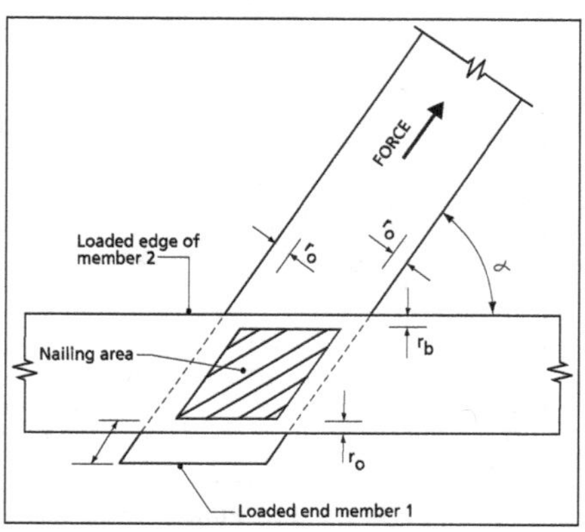

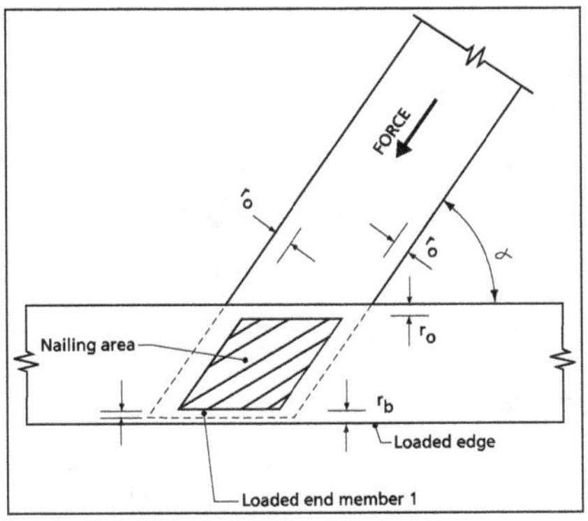

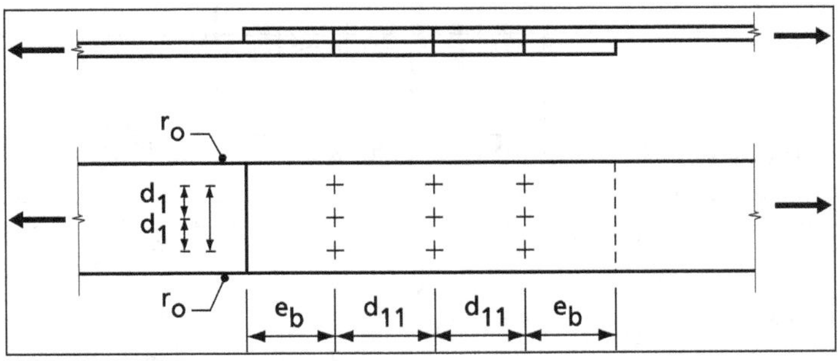

In order to permit the development of the full load at each nail and to avoid splitting the wood, minimum spacing between nails and distances from the edges and ends of the member are necessary. Nailing patterns for use on timber structures are usually available locally. They depend on the quality and type of nails and timbers used, and are based on the safe lateral nail load.

Connections in Steel Structures

Connections may be bolted, riveted or welded. The principal design considerations are shear, tension and compression, and the calculations are relatively straightforward for the types of design covered.

Stability

Stability problems in a building are chiefly the result of horizontal loads, such as those resulting from wind pressure, storage of granular products against walls, soil pressure against foundations and sometimes earthquakes.

Overturning of foundation walls and foundation piers and pads is counteracted by the width of the footing and the weight of the structure. Only in special cases will it be necessary to give extra support in the form of buttresses.

Overturning of external walls is counteracted by the support of perpendicular walls and partitions. However, that not all types of wall, for example framed walls, are adequately rigid along their length without diagonal bracing. If supporting walls are widely spaced and/or the horizontal loads are large, extra support can be supplied by the construction of piers, columns or buttresses.

Diagonal bracing is used to make framed walls and structures stiff. Long braces should preferably transfer the load with a tensile stress to avoid buckling. Braces are usually supplied in pairs, i.e. on both diagonals, so that one will always be in tension independently of the wind direction.

Table: Basic lateral loads per nail.

Continental nail diameter (mm)			2.1	2.4		2.8	3.1		3.4		3.8	4.2		4.6	5.1	5.6		6.1
Kenya nail diameter (mm)	1.8	2.0			2.65			3.35		3.75			4.5			5.6	6.0	
Basic lateral nail load (N)	90	120	140	190	220	250	310	350	370	430	450	540	600	630	750	880	960	100

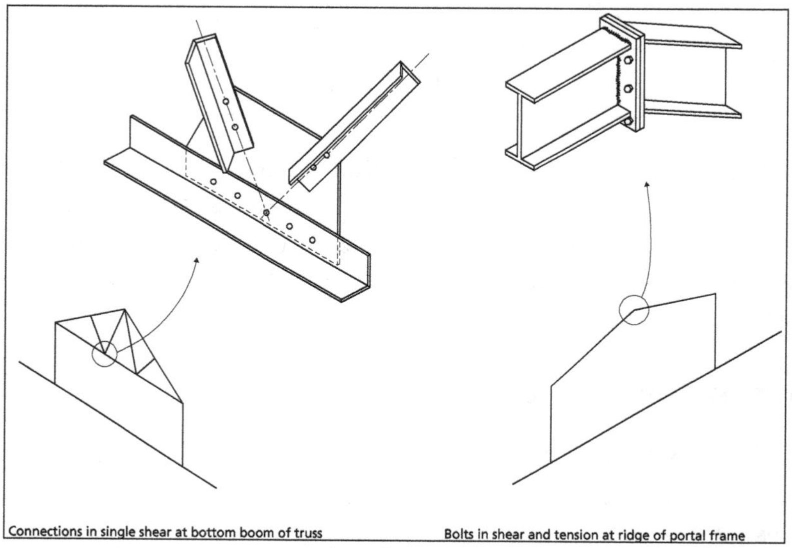

Connections for steel frames.

If the framed wall is covered with a sheet material, such as plywood, chipboard or metal sheets, the lateral forces on the frame can be counteracted by shear in the sheets. This design requires the sheets to be securely fixed to the frame, both horizontally and vertically. The sheets must be strong enough to resist buckling or failure through shear.

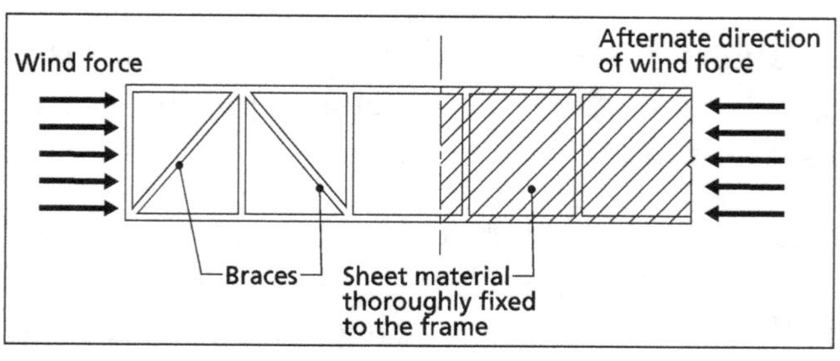

Masonry and concrete walls that are stiff and capable of resisting lateral wind loading are called shear walls. Portal or rigid frame buildings are normally stable laterally, when the wind pressure acts on the long sides. However, when the wind loads occur at the gable ends, the frames may need extra support from longitudinal bracing. Tension rods are frequently used.

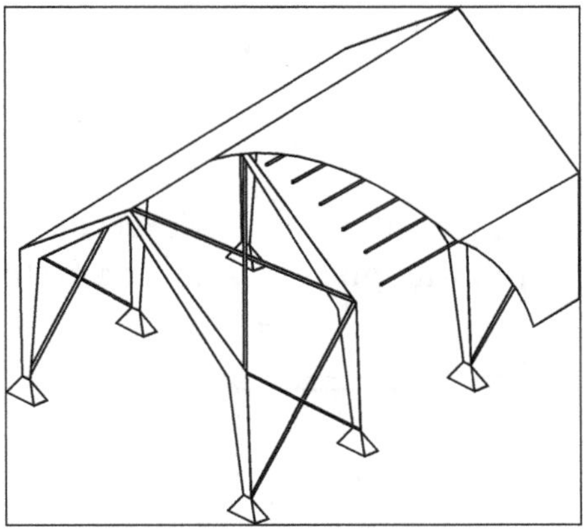

Bracing for portal frame.

Post-and-beam or shed-frame buildings will, in most cases, require wind bracing, both along and across the building because there are no rigid connections at the top of the wall to transfer loads across and along the building. The same applies to buildings employing roof trusses. End bracing should be installed.

Walls with long spans between the supporting cross walls, partitions or buttresses tend to bend inwards under the wind load, or outwards if bulk grain or other produce is stored against the wall. At the bottom of the wall this tendency is counteracted by the rigidity of the foundation (designed not to slide) and the support of a floor structure. The top of the wall is given stability by the support of the ceiling or roof structure, or a specially designed wall beam that is securely anchored to the wall.

The designer must consider the ability of the building to withstand horizontal loading from any and all directions, without unacceptable deformation.

Retaining Walls

Wall Failure

Walls are commonly used to retain soil on sloping sites, water in a pond or bulk products within a storage area. There are several limiting conditions which, if exceeded, can lead to the failure of a retaining wall. Each must be addressed in designing a wall.

1. Overturning: This occurs when the turning moment resulting from lateral forces exceeds that exerted by the self-weight of the wall. The factor of safety against overturning should be at least two.

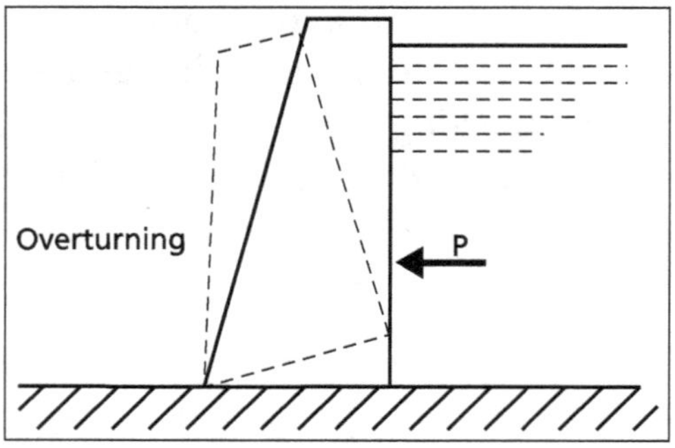

2. Sliding: The wall will slide if the lateral thrust exceeds the frictional resistance developed between the base of the wall and the soil. The factor of safety against sliding should be about two.

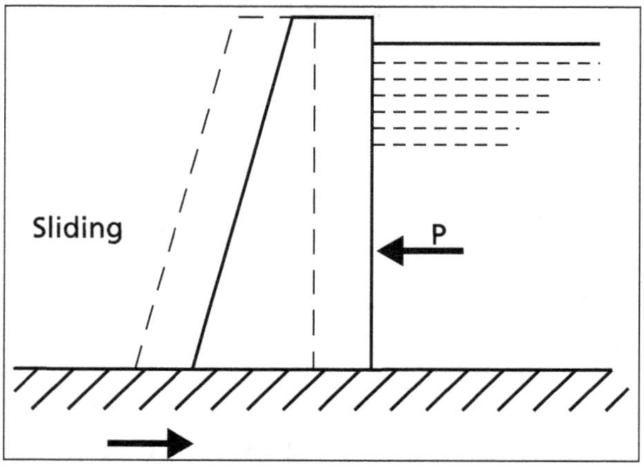

3. Bearing on ground: The normal pressure between the base of the wall and the soil beneath can cause a bearing failure of the soil, if the ultimate bearing capacity is exceeded. Usually the allowable bearing pressure will be one-third of the ultimate value. Note that the pressure distribution across the base is not constant.

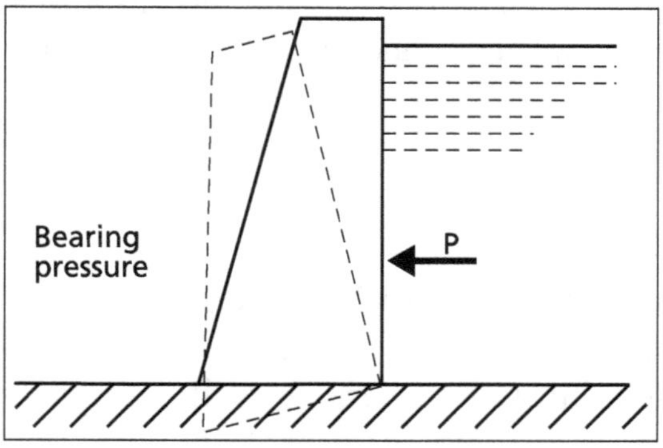

4. Rotational Slip: The wall and a large amount of the retained material rotate about point O if the shear resistance developed along a circular arc is exceeded. The analysis is too complex to include here.

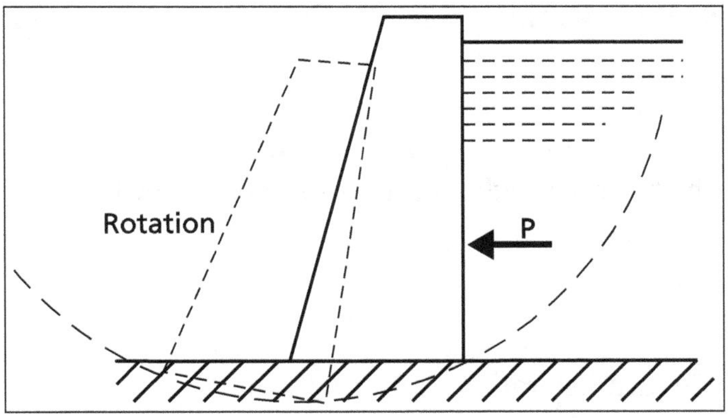

5. Wall material failure: The structure itself must be capable of withstanding the internal stresses set up, that is to say, the stresses must not exceed allowable values. Factors of safety used here depend on the material and the level of the designer's knowledge of the loads actually applied. Naturally, both shear and bending must be considered, but the most critical condition is likely to be tension failure of the 'front' facet.

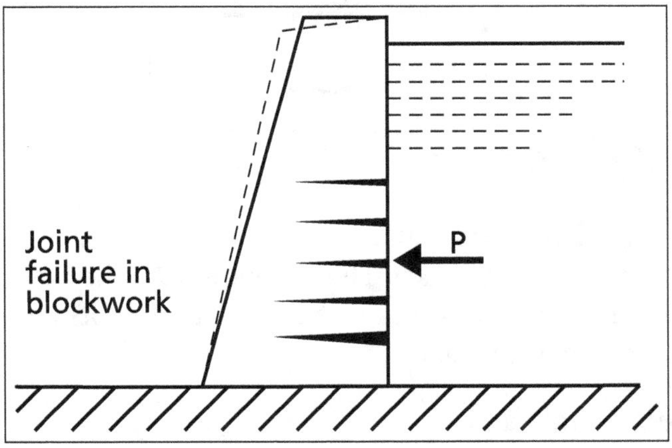

Gravity walls and dams are dependent on the effect of gravity, largely from the self-weight of the wall itself, for stability. Other types of wall rely on a rigid base, combined with a wall designed against bending, to provide an adequate structure.

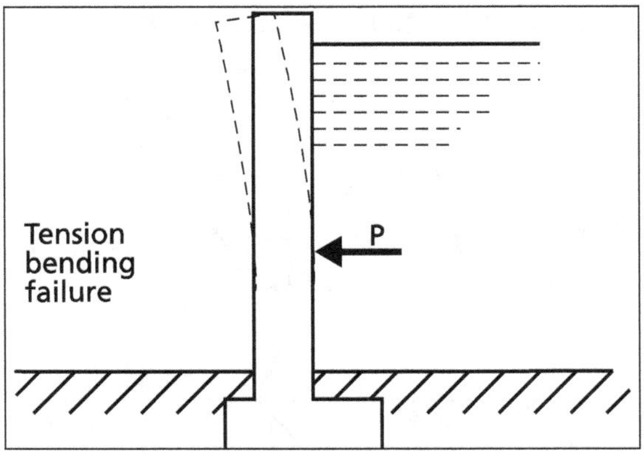

Pressure Exerted by Retained Material

Liquid Pressure

The pressure in a liquid is directly proportional to both the depth and the specific weight of the liquid (w) which is the weight per unit volume, $w = \rho g$ (N/m³),

Where,

- ρ = density of liquid (kg/m³).
- g = gravitational acceleration (9.81 m/s²).

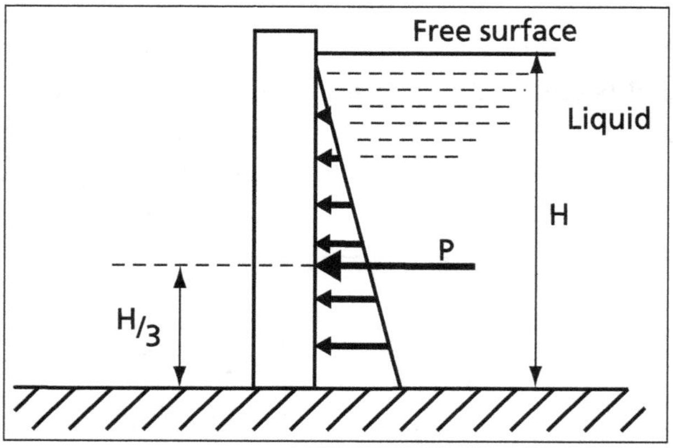

The pressure at a given depth acts equally in all directions, and the resultant force on a dam or wall face is normal to the face. The pressure from the liquid can be treated as a distributed load with linear variation in a triangular load form, with a centroid two-thirds of the way down the wet face.

$$p = \rho g H = wH \,(\mathrm{N/m^2}) \text{ and}$$

$$P = \frac{wH^2}{2} \text{ acting at a depth of } \frac{2}{3} H$$

It should be noted that a wall retaining a material that is saturated (waterlogged) must resist this liquid pressure in addition to the lateral pressure from the retained material.

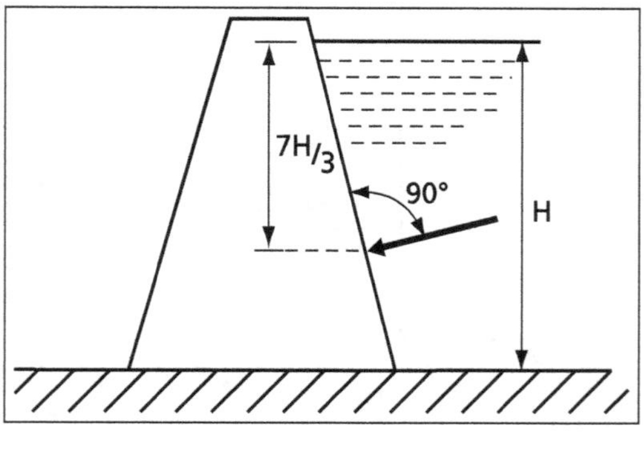

Example: Design of a gravity wall retaining water.

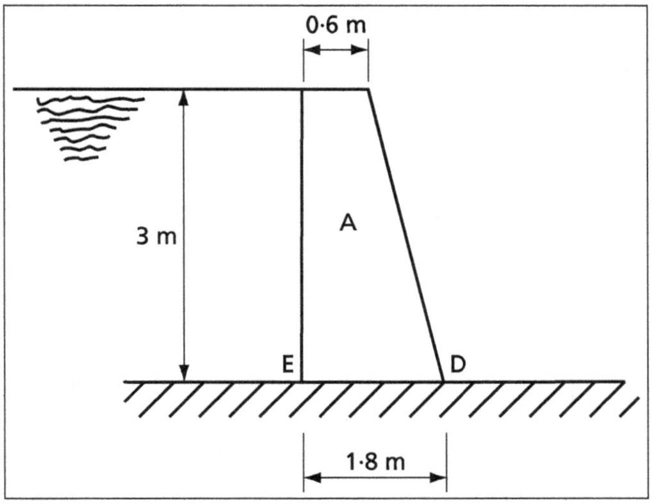

Consider a mass concrete dam with the cross-section shown, which retains water to 3 m depth.

Assume:

- Ground safe bearing capacity: 300 kPa.
- Coefficient of sliding friction at base: 0.7.
- Specific weight of concrete: 23 kN/m³.

1. Find water force P: All calculations per metre length of wall,

$$P = \frac{wH^2}{2} = \frac{9.8 \times 10^3\ 3^2}{2} = 44.1 \text{ kN}$$

2. Find mass of one metre length of wall:

$$\begin{aligned} W &= A \times \text{specfic weight} \\ &= 3 \times \frac{(0.6+1.8)}{2} \times 23 = 82.8 \text{kN} \end{aligned}$$

3. Find line of action of w: Taking moments of area about vertical face:

$$\begin{aligned} \overline{X} &= \frac{A_1X_1 + A_2X_2}{A_1 + A_2} \\ &= \frac{(3 \times 0.6 \times 0.3) + (0.6 \times 3 \times 1.0)}{1.8 + 1.8} = 0.65 \text{ m} \end{aligned}$$

Hence the self-weight of the wall acts 0.25 m to left of the base centre line.

4. Find the vertical compressive stress on the base:

$$P_c = \frac{W}{A} = \frac{82.8}{1 \times 1.8} = 46 \text{ kPa}$$

5. Find the moment about the centre line of the base:

$$M = (1 \times 44.1) - (0.25 \times 82.8);\ (\text{clockwise}) - (\text{anticlockwise})$$
$$M = 23.4 \text{ kNm}$$

6. Find the bending stresses/pressures:

$$\sigma_b = P_b = \frac{MI}{Y_{\max}}$$

Where,

$$I = \frac{bd^3}{12} = \frac{1 \times 1.8^3}{12} = 0.486\,\text{m}^4$$

$$Y_{\max} = \pm\frac{1.8}{2} = \pm 0.9\,\text{m}$$

$$\sigma_b = P_b = \pm\frac{23.4 \times 0.486}{0.9} = 12.6 \text{ kPa}$$

7. Find the actual stresses/pressures:

$$\sigma = P = \frac{W}{A} + \frac{M_y}{I}$$

$$\sigma_E = P_E = 46 + 12.6 = 58.6 \text{ kPa (compression)}$$
$$\sigma_D = P_D = 46 - 12.6 = 33.4 \text{ kPa (compression)}$$

(Compression only indicates the resultant *P*, and *W* would intersect the base line within its middle third).

8. Compare maximum pressure with allowable bearing capacity:

$$P_{\ddot{u}} = 58.6 \text{ kPa}$$

This is less than the allowable safe bearing capacity of the soil. Hence the wall-soil interface is safe against bearing failure.

9. Compare actual stresses in the wall with allowable values:

Maximum stress = 58.6 kPa (compression) and no tensile stress at any point across wall. Hence the wall material is safe.

10. Check overturning:

Overturning moment about D = 44.1 × 1 = 44.1 kNm

Stabilising moment about D = 82.8 × 1.15 = 95.22 kNm

Factor of safety against overturning = 94.22 / 44.1 = 2.16

The wall is safe against overturning.

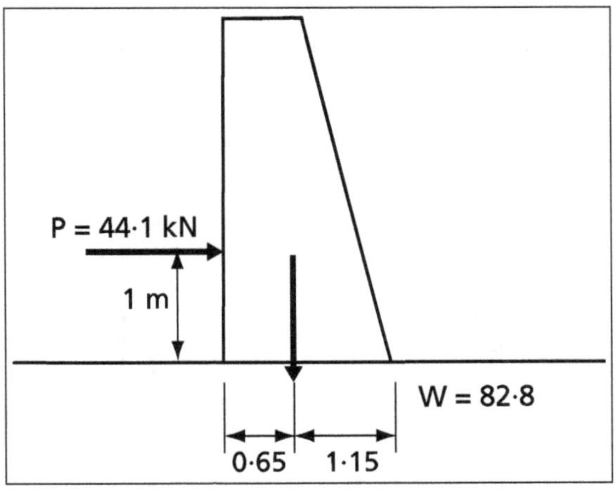

11. Check sliding:

Frictional resistance = μW

μW= 0.7 × 82.8 = 58 kN

Horizontal thrust = *P* = 44.1 kN

As the required factor against sliding is 2, there is a deficiency of (2 × 44.1) - 58 = 30.2 kN.

Additional anchorage against sliding should be provided.

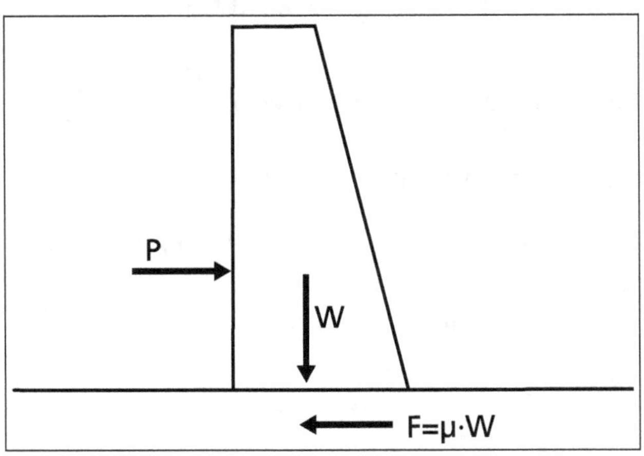

Example: Design a circular water tank with the following dimensions/properties:

- Diameter 5 m, depth of water 3 m
- Water weighs 9.8 × 103 N/m3
- Pressure (*P*) at a depth of 3 m

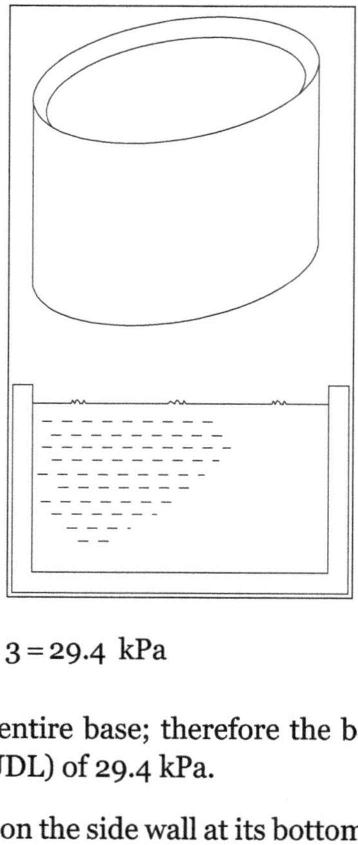

$$P_3 = wH = 9.8 \times 10^3 \times 3 = 29.4 \text{ kPa}$$

This acts vertically over the entire base; therefore the base should be designed for a uniformly distributed load (UDL) of 29.4 kPa.

Pressure P_3 also acts laterally on the side wall at its bottom edge. This pressure decreases linearly to zero at the water surface.

Total force on base = $P_3 A_B = 29.4 \times \frac{(\pi \times 5^2)}{4} = 577.3$ kN

Total force on the side per metre of perimeter wall: $\frac{P_3 H}{2} = \frac{29.4 \times 3}{2} = 44.1$ kN/m run.

Pressure Exerted by Granular Materials

Granular materials such as sandy soils, gravelly soils and grain possess the property of internal friction (friction between adjacent grains), but are assumed not to possess the property of cohesion. If a quantity of such material in a dry condition is tipped on to a flat surface, it will form a conical heap: the shape maintained by this internal friction between grains. The angle of the sloping side is known as the angle of repose.

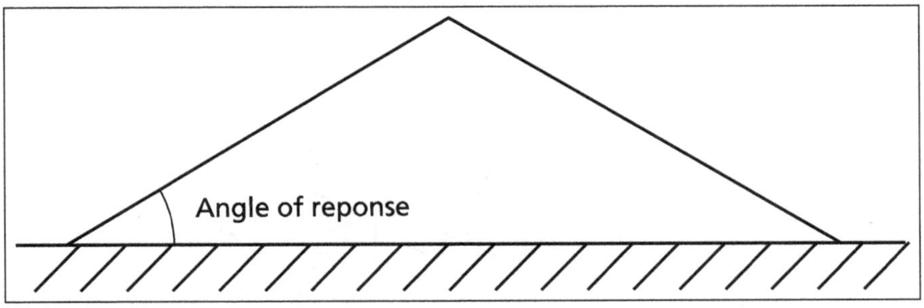

For a dry material, the angle of repose is usually equal to the angle of shearing resistance of the material. This angle of shearing resistance is called the angle of internal friction (θ). The angle of friction is the essential property of a granular material on which Rankine's theory is based. This theory enables the lateral pressure to be expressed as a proportion of the vertical pressure, which was shown to depend on specific weight and depth only.

In this case, at a depth h the active lateral pressure is given by:

$$P = k \times w \times h$$

Where,

k = a constant dependent on the materials involved.

Although there is some friction between the retained material and the wall face, usually this is disregarded, giving a relatively simple relationship for k:

$$k = \frac{1 - sin\theta}{1 + sin\theta}$$

Where, θ = the angle of friction.

$$p_a = \frac{1 - sin\theta}{1 + sin\theta} \times wH \, (\mathrm{N/m^2})$$

Where,

p_a = total force per metre of wall-face (N)

$$P_a = \frac{1 - sin\theta}{1 + sin\theta} \times \frac{wH^2}{2} \text{ (N/m length of wall)}$$

P_a = total force per metre of wall face (N)

This gives the approximate horizontal resultant force on a vertical wall face when it is retaining material that is level with the top of the wall. If the surface of the retained material is sloping up from the wall at an angle equal to its angle of repose, a modification is required.

Example: Wall retaining soil.

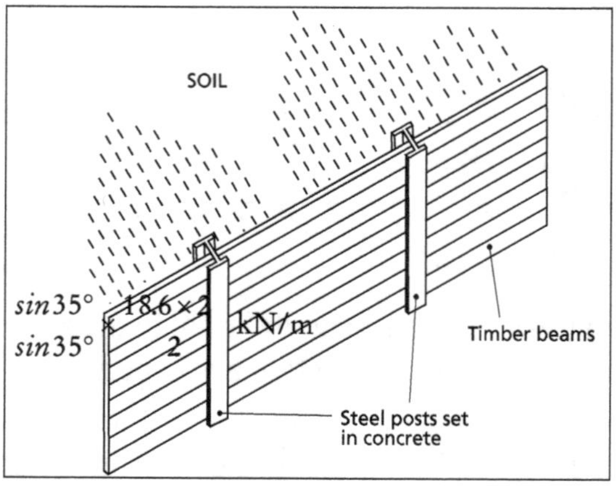

Consider the wall shown retaining loose sandy soil to a depth of 2 metres. Tables provide angle of friction equal to 35° and specific weight equal to 18.6 kN/m³. Assuming a smooth vertical surface and horizontal soil surface, Rankine's theory gives:

$$P = \frac{1 - sin\theta}{1 + sin\theta} \times \frac{wH^2}{2}$$

$$P = \frac{1 - sin\,35°}{1 + sin\,35°} \times \frac{18.6 \times 2^2}{2} \text{ kN/m length of wall}$$

$$P = 10.1 \text{ kN/m length of wall.}$$

If steel posts are placed at 2.5 m centres, each post can be approximated to a vertical cantilever beam 2.5 m long, carrying a total distributed load of 10.1 × 2.5 = 25.25 kN of linear variation from zero at the top to a maximum at the base. The steel post and foundation concrete must be capable of resisting the applied load, principally in bending but also in shear.

The timber crossbeams can be analysed as beams simply supported over a span of 2.5 m, each carrying a uniformly distributed load. This load is equal to the product of the face area of the beam and the pressure in the soil at a depth indicated by the centroid of the area of the beam face.

$$P_a = \frac{1 - sin\theta}{1 + sin\theta} \times wh$$

if beam face is 0.3 m high,

h = 2.0 – 0.15= 1.85 m

P = 0.27 × 18.6 × 1.85 = 9.29 kN/m²

Total uniformly distributed load on the beam = 9.29 × 0.3 × 2.5 = 6.97 kN

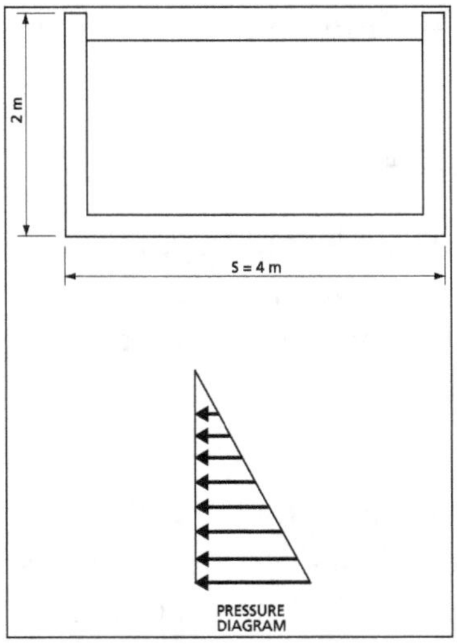

The maximum bending moment at the centre of the span can be determined and the beam section checked.

Example: Grain storage bin (The theory given does not apply to deep bins). A shallow bin can be defined as one with a sidewall height of less than:

$$\frac{S}{2}\tan\left(45°+\frac{\theta}{2}\right) \text{ for a square bin of side length S.}$$

Consider a square bin of side length 4 metres retaining shelled maize/corn to a depth of 2 metres. Assume θ = 27°; specific weight is 7.7 kN/m³.

Critical height is:

$$\frac{S}{ü}\tan\left(45°+\frac{\theta}{\ }\right)=\frac{4}{\ }\tan(45°+13.5°)=2\times1.63=3.26 \text{ m}$$

Design in the same way as the shallow bin because the depth of grain is only 2 metres.

Maximum pressure at the base of the wall:

$$P_a=\frac{1-sin\theta}{1+sin\theta}\times wH$$

$$=\frac{1-sin\,27°}{1+sin\,27°}\times7.7\times2=5.78 \text{ kN/m}^2$$

or resultant force $P=\frac{5.78\times2^2}{2}=11.56 \text{ kN/m}$

(acting 2 /3 m above the base of the wall).

Note that the design of the wall is complex if it consists of a plate of uniform thickness, but if the wall is thought of as comprising a number of vertical members cantilevered from the floor, an approach similar to that for the soil-retaining wall can be used.

Designing for Earthquakes

In areas where earthquakes occur frequently, buildings must be designed to resist the stresses caused by tremors. While the intensity of tremors can be much greater in loosely compacted soil than in firm soil or solid bedrock, one- and two-storey buildings are at greater risk on firm ground or bedrock because of the shorter resonance periods.

Casualties are most likely to be caused by the collapse of walls causing the roof to fall, and the failure of projecting elements such as parapets, water tanks, non-monolithic chimneys and loose roof coverings. Outbreaks of fire caused by fractures in chimneys or breaks in mains supply lines present an additional hazard.

While small buildings with timber frame walls, or a wooden ring beam supported by the posts of a mud-and-pole wall, can resist quite violent earthquakes, the following measures will increase the resistance of a large building to collapse from earth tremors:

- Use a round or rectangular shape for the building. Other shapes such as 'L' 'T' or 'U' should be divided into separate units. To be effective, this separation must be carried down through to the foundation.
- Avoid large spans, greatly elongated walls, vault-and-dome construction and wall openings in excess of one-third of the total wall area.
- Construct a continuously reinforced footing that rests on uniform soil at a uniform depth – even on sloping ground.
- Fix the roof securely, either to a continuously reinforced ring beam on top of the walls, or to independent supports, which will not fail even if the walls collapse.
- Avoid projecting elements, brittle materials and heavy materials on weak supports.
- Avoid combustible materials near chimneys and power lines.

Ductile structures have many joints that can move slightly without failing, e.g. bolted trusses. Such structures have a greater capacity to absorb the energy of earthquake waves. A symmetrical, uniformly distributed ductile framework with the walls securely fixed to the frame is suitable for large buildings. Masonry walls are sensitive to earthquake loads and tend to crack through the joints. It is therefore important to use a good mortar and occasionally reinforcing will be required.

Structural Engineering

Structural Engineering is a specialty within Civil Engineering. Structural Engineers create drawings and specifications, perform calculations, review the work of other engineers, write reports and evaluations, and observe construction sites. A Professional Engineer's license is required in order to practice Structural Engineering. A license can be obtained only after completing a prescribed amount of education and work experience, and taking a 2-day exam. In California and other states, certain structures, such as hospitals and schools, require a Structural Engineer's license, which can be obtained after 3 years additional experience and taking another exam.

Analyzes and Designs

The basic tasks of structural engineering relate numerical quantities of physical forces to physical configurations of force-resisting elements. Analysis is the process of determining forces in each element in a structure (such as a beam) when the configuration of elements is already defined. Design is the process of configuring elements to resist forces whose values are already known. Analysis and Design are complementary procedures in the overall process of designing new structures. After performing a preliminary design, the designer estimates the final configuration of elements of a structure, but only until an analysis is performed can the forces in those elements be known. After performing an analysis, the element forces are known, and the elements can be designed (their configuration can be chosen) more precisely. The process iterates between analysis and design until convergence is achieved.

Gravity Support and Lateral Force Resistance

Structures are subject to vertical, or "Gravity" Loads and horizontal, or "Lateral" Loads. Gravity loads include "dead", or permanent, load, which is the weight of the structure, including its walls, floors, finishes, and mechanical systems, and "live", or temporary load, which is the weight of a structure's contents and occupants, including the weight of snow. Lateral loads include those generated by the wind, earthquakes, or explosions. Structural elements must be designed so that, as a system, the structure can resist all loads that will act upon it.

Buildings, Bridges and other Structures

Structures are any system that resists vertical or horizontal loads. Structures include large items such as skyscrapers, bridges, and dams, as well as small items such as bookshelves, chairs, and windows. Most every day "structures" are "designed" by testing, or trial and error; while large, unique, or expensive structures that are not easily tested are generally designed by a qualified structural engineer using mathematical calculations. Most practicing structural engineers design and analyze buildings, bridges, power plants, electrical towers, dams, and other large structures that are essential to life as we know it.

References

- Structure-definition: designingbuildings.co.uk, Retrieved 11 April 2019
- Building, 12-basic-components-building-structure-34024: theconstructor.org, Retrieved 14 June 2019
- Whatisase: seaoc.org, Retrieved 18 April 2019
- Types-of-structures: excelyourcivil.blogspot.com, Retrieved 25 August 2019
- Structural-idealization: civilengineeringforum.me, Retrieved 14 July 2019

2

Structural Elements

Structural elements are used in structural analysis for splitting of a complex structure into simple elements. It includes beams, trusses, columns, cables, arches, etc. The topics elaborated in this chapter will help in gaining a better perspective about these structural elements.

Columns

A column can be defined as a vertical structural member designed to transmit a compressive load. A Column transmits the load from ceiling/roof slab and beam, including its own weight to the foundation. Hence it should be realized that the failure of a column results in the collapse of the entire structure. The design of a column should therefore receive importance.

In the modern construction industry, Columns are mostly constructed by concrete; apart from that materials such as Wood, Steel, Fibre-reinforced polymer, Cellular PVC, and Aluminium too are been used. The type of material is been decided on the scale, coast and application of the construction.

Types of Columns in Construction

There are several types of columns which are used in different parts of structures. Column is vertical structural members that carry loads mainly in compression. It might transfer loads from a ceiling, floor slab, roof slab, or from a beam, to a floor or foundations.

Commonly, columns also carry bending moments about one or both of the cross-section axes.

Types of Columns in Building Construction

Columns are classified based on the several conditions which include:

- Based on Types of Reinforcement.
- Based on Types of Loading.

- Based on Slenderness Ratio.
- Based on Shape.
- Based on Construction Material.

Based on Types of Reinforcement

Tied Column

This type of column is commonly construction from reinforced concrete. Longitudinal reinforcement are confined within closely spaced tie reinforcement. It is estimated that 95% of all columns in buildings are tied.

Spiral Column

Spiral column is also construction from reinforced concrete. In this type of column, longitudinal bars are confined within closely spaced and continuously wound spiral reinforcement.

Spiral reinforcement provides lateral restrains (Poisson's effect) and delays axial load failure (ductile).

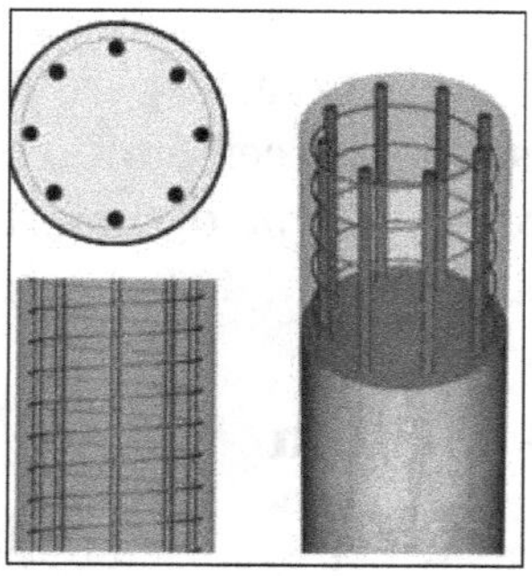

Spiral Column.

Composite Column

When the longitudinal reinforcement is in the form of structural steel section or pipe with or without longitudinal bars, it is called as a composite column.

These types of column have high strength with fairly small cross-section, in addition to exhibit good fire performance.

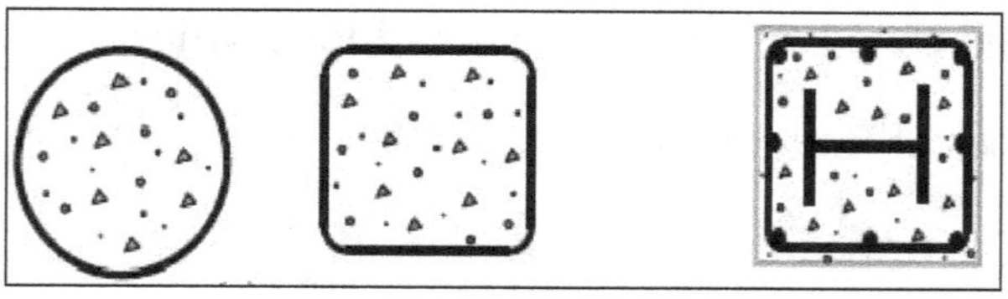

Composite column.

Based on Types of Loading

Axially Loaded Column

If vertical axial loads act on the center of gravity of the cross-section of the column, then it is termed as axially loaded column.

Axially loaded column is rare in construction since coinciding vertical loads on the center of gravity of column cross-section is not practical.

Interior column of multi-storey buildings with symmetrical loads from floor slabs from all sides is an example of this type of column.

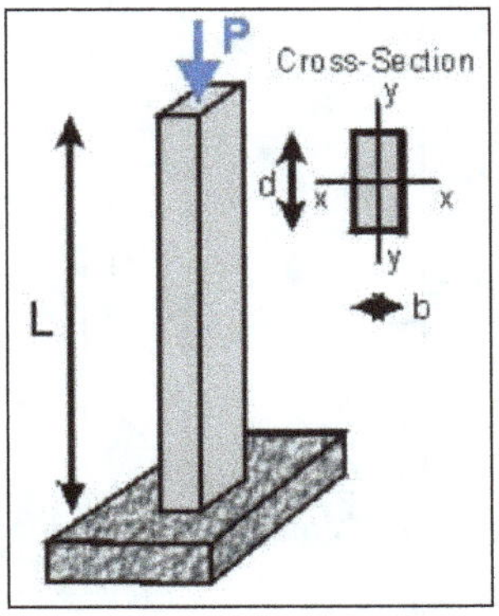

Axially loaded column.

Column with Uniaxial Eccentric Loading

When vertical loads do not coincide with center of gravity of column cross-section, but rather act eccentrically either on X or Y axis of the column cross-section, then it is called uniaxially eccentric loading column.

Column with uniaxial loading are generally encountered in the case of columns rigidly connected beam from one side only such as edge columns.

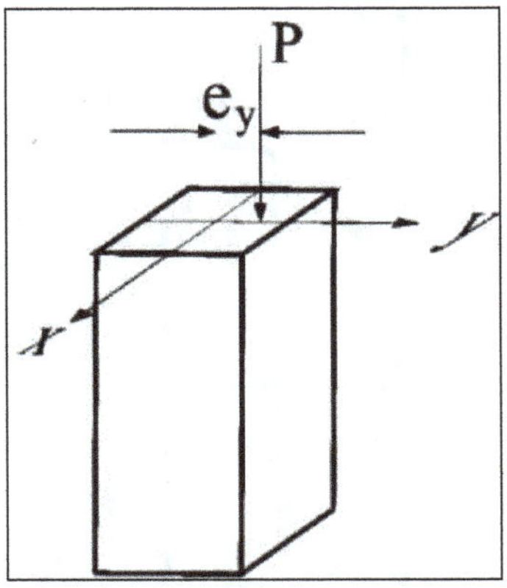

Column with uniaxial eccentric loading.

Column with Biaxial Eccentric Loading

When vertical on the column is not coincide with center of gravity of column cross-section and does not act on either axis (X and Y axis), then the column is called biaxially eccentric loaded column.

Columns with biaxial loading are common in corner columns with beams rigidly connected at right angles at the top of columns.

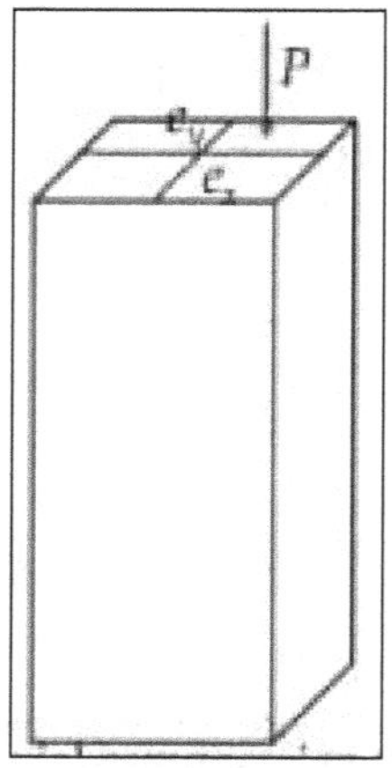

Column with biaxial eccentric loading.

Based on Slenderness Ratio

Based on slenderness ratio, (effective length/ least lateral dimension), columns are categorized as follow.

Short Column

If the ratio effective length of the column to the least lateral dimension is less than 12, the column is called as the short column. A short column fails by crushing (pure compression failure).

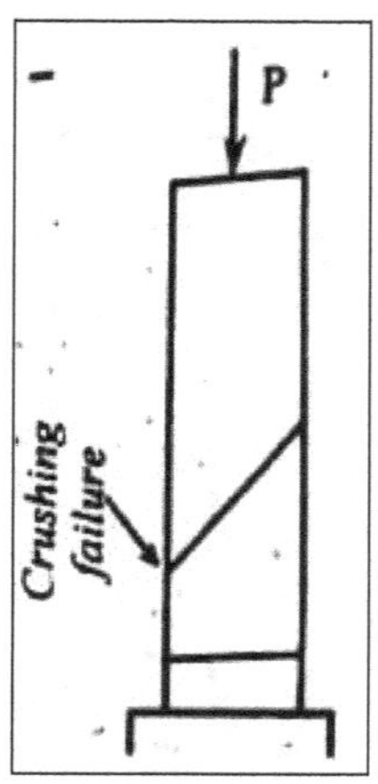

Short column.

Long Column

If the ratio effective length of the column to the least lateral dimension exceeds 12, it is called as long column. A long column fails by bending or buckling.

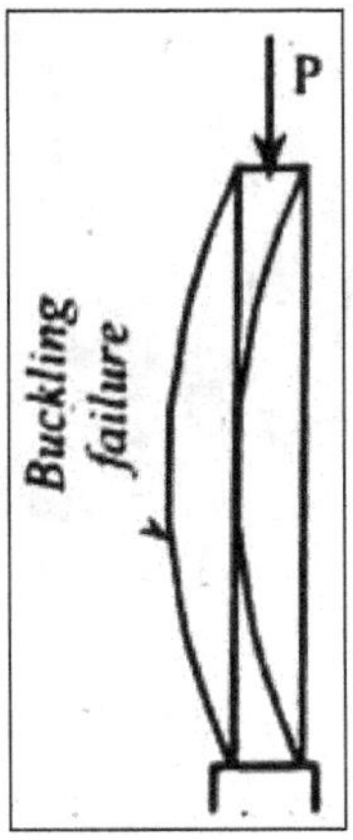

Long column.

Based on Shape

Square or Rectangular Column

They are generally used in the construction of buildings. It is much easier to construct and cast rectangular or square columns than circular ones because of ease of shuttering and to support it from collapsing due to pressure while the concrete is still in flowable form.

Square column.

Circular Column

They are specially designed columns, which are mostly used in piling and elevation of the buildings.

Circular column.

L-Shape Column

Commonly, L-shaped column is utilized in the corners of the boundary wall and has similar characteristics of a rectangular or square column.

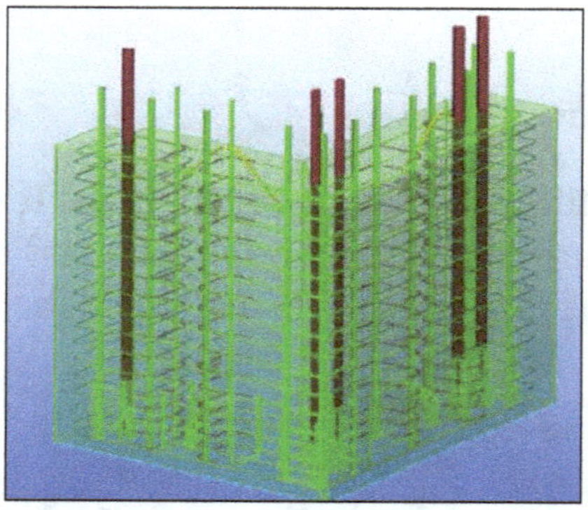

L-shaped column.

T-Shape column

It is utilized based on design requirements of a structure. T-Shaped column is widely used in the construction of bridges.

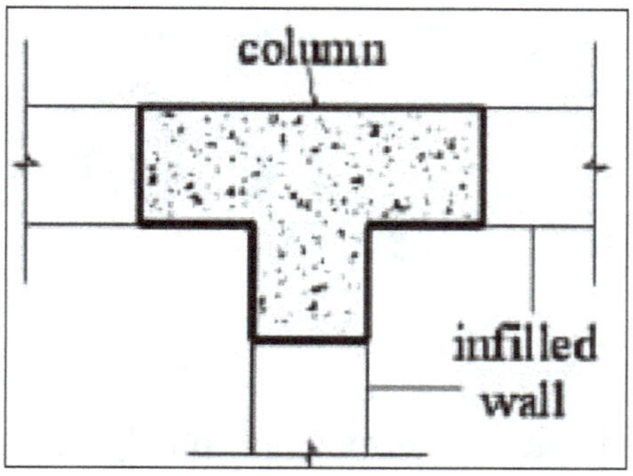

T-shaped column.

Shape of Steel Column

There are different standard and built up shape of steel columns which are shown in the figure. Common shapes of steel columns include I, channel, equal angle, and T-shape.

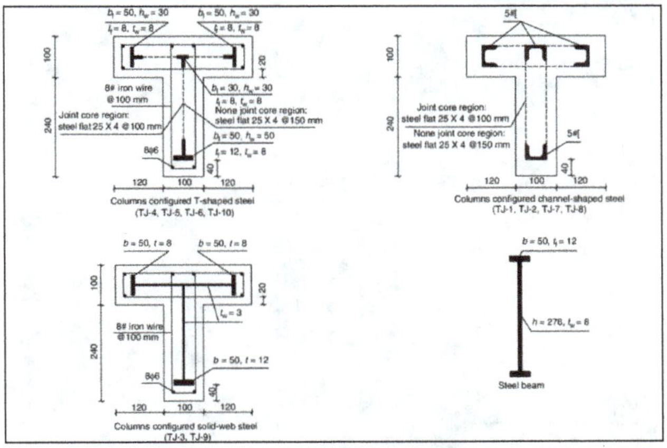

Steel column cross-section shape.

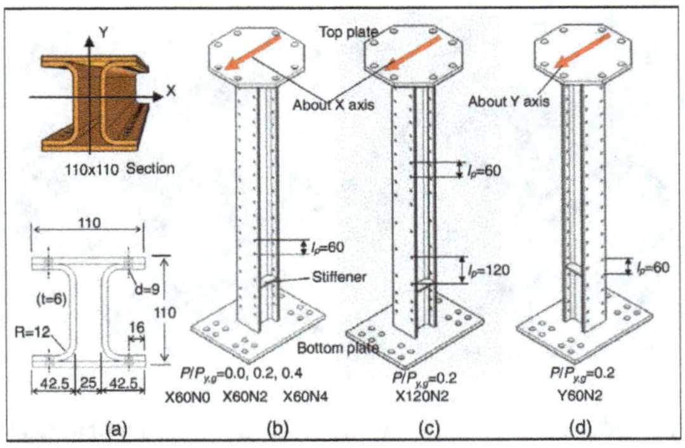

Steel column cross-section shape (built up).

Shape of Composite Column

The usual shapes of composite columns are shown in figure.

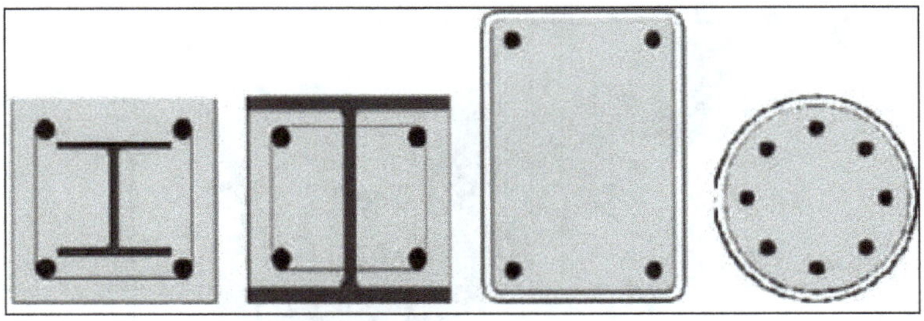

Composite column shape.

Based on Construction Material

Types of columns based on construction materials include: Reinforced Concrete, Steel, timber, Brick, Block, and Stone Column.

Types of Column; A-reinforced concrete, B- steel, C-timber, D-brick, E-block, and F-stone.

Beams

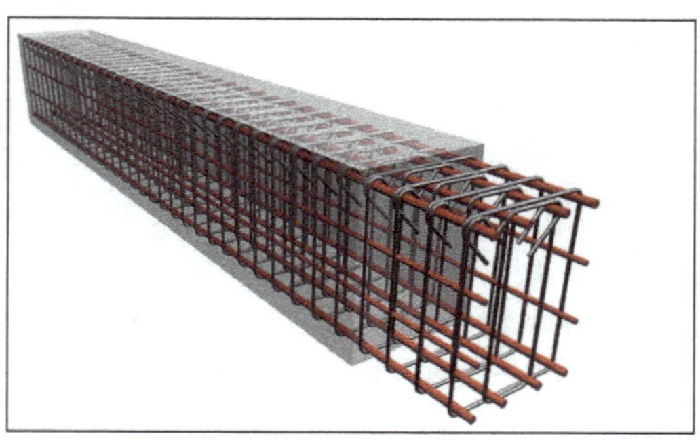

A beam is a horizontal structural member in a building to resist the lateral loads applied to the beams axis. The structural member which resists the forces laterally or transversely applied to the (beam) axis is called beam.

In it, the loads are acting transversely to the longitudinal axis, which produces the shear forces and bending moment. The lateral load acting on beams are the main cause bending of the beam. They are responsible to transfer a load from the slab to the column. The load distribution pattern is:

Slab |> Beam |> Column |> Foundation

That beam which is connected with the column and this connection is called direct support while the beam connected with beam and this connection is called indirect support.

Purpose of Beams

It is a structural element that is capable of withstanding load primarily by resisting its bending forces. They are made of steel or reinforced concrete (RCC)or steel. We use it in structure to:

- Resist loads.
- Counter bending moment and shear forces.
- Connect the structure together.
- Provide a uniform distribution of loads.

In the beam, the different reinforcements were used having different purposes such as:

- Support bars: These bars are located in the top portion of the beam and just function to hold the stirrups in place.
- Main Bars: Provide to carry loads (Moments).
- Stirrups: To counter the shear stresses (shear force).

They are generally characterized by their profile (shape, cross-section, length, material). This member of RCC structures is placed horizontally to carry the load and counter both bending and shear stresses.

The standard size of the beams:

- In a residential building is 9" × 12" or 225 mm × 300 mm.
- The minimum size of the RCC beam should not be less than the 9"× 9" or 225mm × 225mm with the addition of slab thickness which is 125mm.

Purpose of Stirrups in the Beam

- Stirrups are used to counter the shear force. It is also called shear reinforcement in the beam.
- Shear force is maximum at the end supports (simply supported beams)

and zero at the mid of the span that's why the spacing of stirrups or rings is closed to each end supports as compared to the mid.

- Stirrups are made in a rectangular shape with reinforcement bars and which is wrapped around the top and bottom bars of the beam.
- Sometimes, stirrups are placed diagonally and vertically to avoid the shear failure in case of cracks in beams.

Types of Beams

Different types of beams are used in construction of building and structures. These are horizontal structural element that withstands vertical loads, shear forces and bending moments. Beams transfer loads imposed along their length to their end points to walls, columns, foundations, etc.

Types of Beams in Constructions

There are different types of beams which are classified based on the following conditions:

- Based on Support Conditions.
- Based on Construction Materials.
- Based on Cross-Section Shapes.
- Based on Geometry.
- Based on Equilibrium Condition.
- Based on Method of Construction.
- Others.

Based on Support Conditions

Simply Supported Beam

It is the one of the simplest structural elements that both ends are rest on supports but are free to rotate. It contains pinned support at one end and roller support at the other end. On the basis of assign load, it sustains shearing and bending.

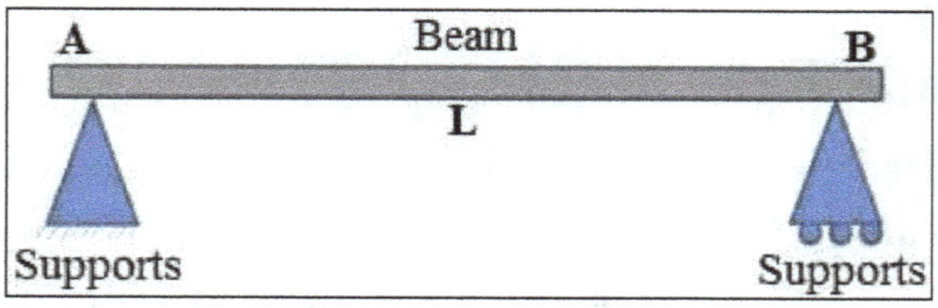

Simply supported beam.

Fixed Beam

It is supported at both ends and fixed to resist rotation. It is also called a built-in beam. The fixed ends produce fixing moments other than the reactions.

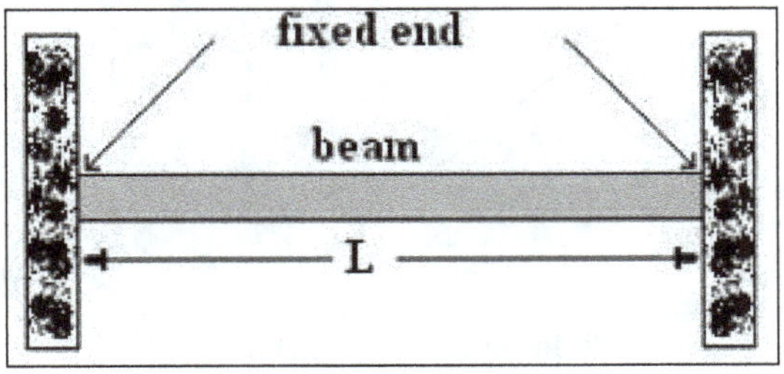

Fixed beam.

Cantilever Beam

If a beam is fixed at one end and set to be free at the end, it is termed as cantilever beam. The beam distributes the load back to the support where it is forced against with a moment and shear stress. Cantilever beams allow the creation of a bay window, balconies, and some bridges.

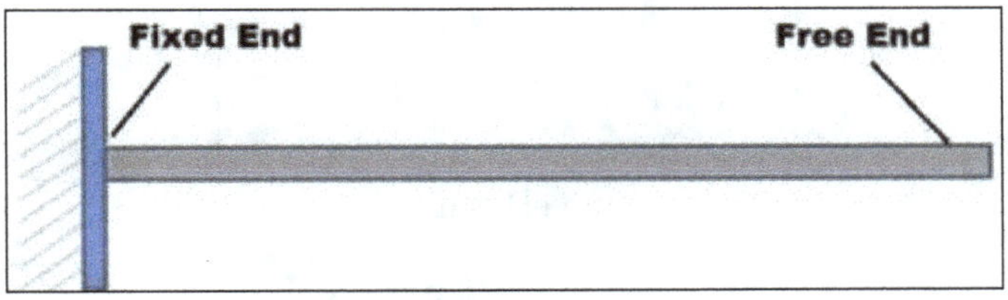

Cantilever beam.

Continuous Beam

A continuous beam has more than two supports distributed along its entire length.

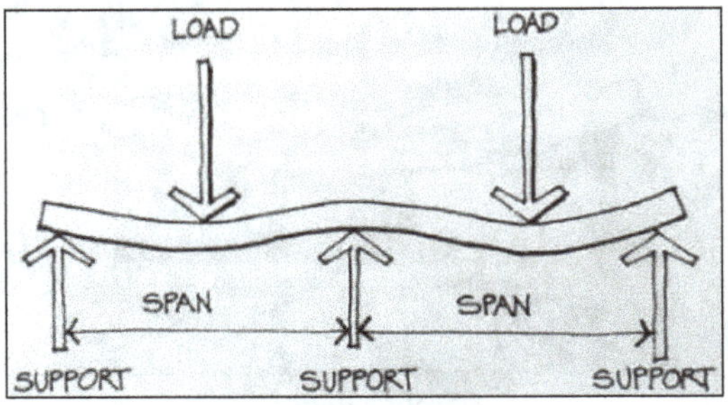

Continuous beam.

Based on Construction Materials

Reinforced Concrete Beams

It is constructed from concrete and reinforcement as shown in figure.

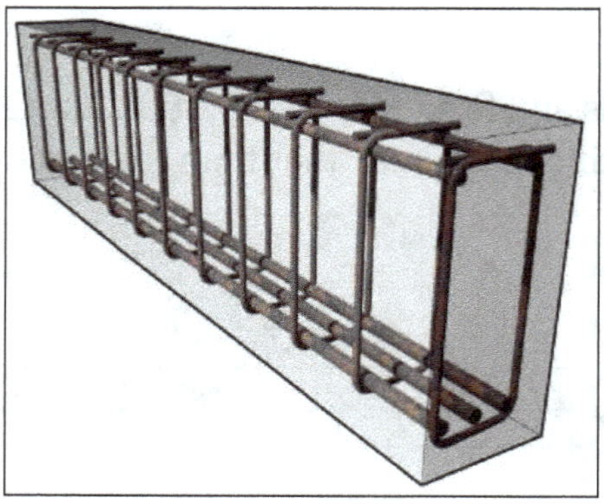

Reinforced concrete beam.

Steel Beams

It is constructed from steels and used in several applications.

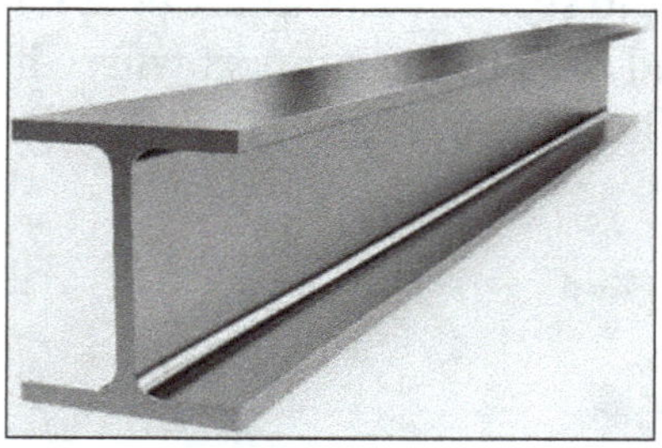

Steel beam.

Timber Beams

This type of beam is constructed from timber and used in the past, but its application is significantly declined now.

Timber Beam.

Composite Beams

Composite beams are constructed from two or more different types of materials such as steel and concrete, and various valid cross-sections have been utilized as shown in figure.

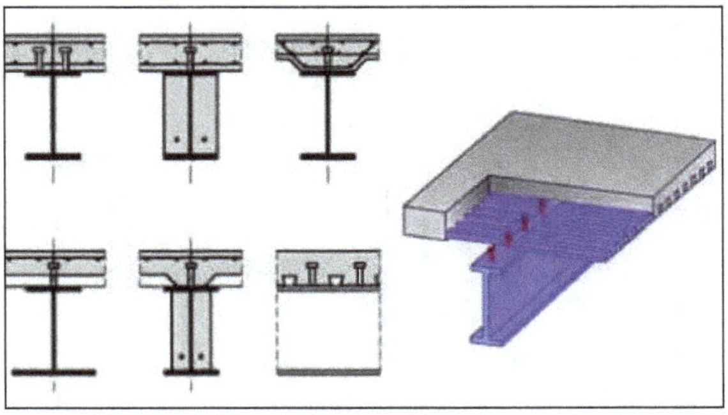

Composite beam.

Based on Cross-Section Shapes

Several cross-sectional shapes of beams are available and used in different parts of of structures. These beams can be constructed from reinforced concrete, steel, or composite materials.

Reinforced concrete cross-sectional shapes include:

Rectangular Beam

This type of beam is widely used in the construction of reinforced concrete buildings and other structures.

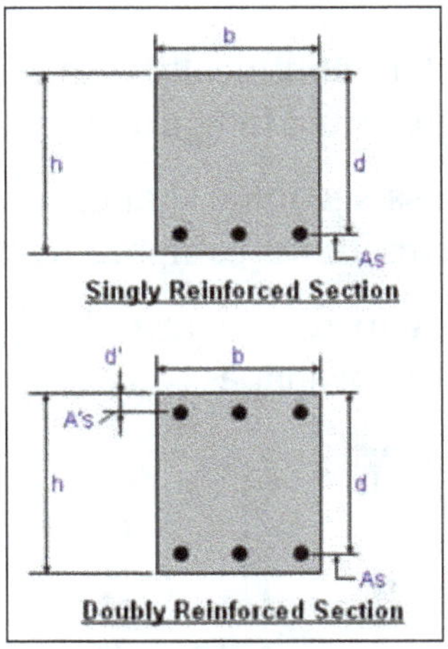

Rectangular Reinforced concrete beam.

T-Section Beam

This type of beam is mostly constructed monolithically with reinforced concrete slab. Sometimes, Isolated T-beam are constructed to increase the compression strength of concrete.

Added to that, inverted T-beam can also be constructed according to the requirements of loading imposed.

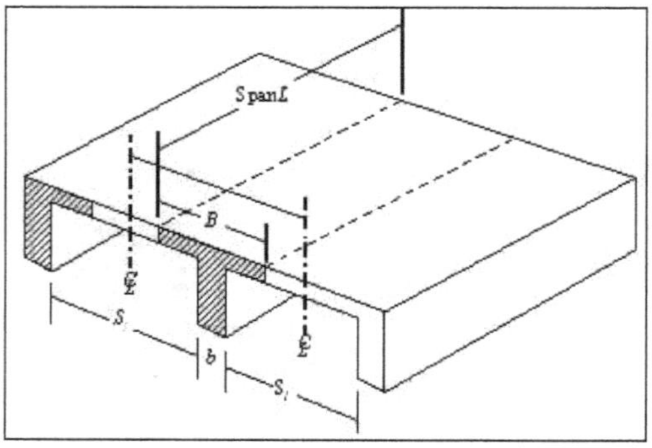

T-beam.

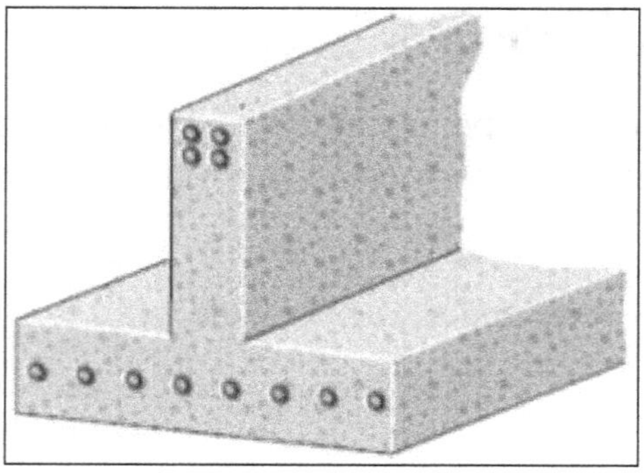

Inverted T-beam.

L-Section Beam

This type of beam is constructed monolithically with reinforced concrete slab at the perimeter of the structure, as illustrated in figure.

There are various steel beam cross-sectional shapes. Each cross-sectional shape offer superior advantages in a given conditions compare with other shapes.

Square, rectangular, circular, I-shaped, T-shaped, H-shaped, C-shaped, and tubular are examples of beam cross-sectional shapes constructed from steel.

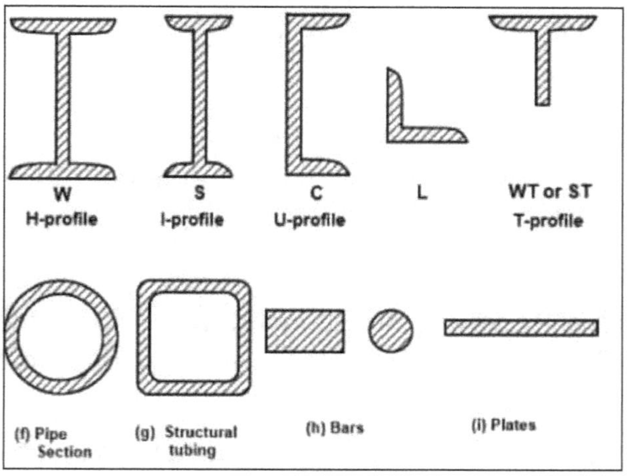

Steel beam cross-sectional shapes.

Based on Geometry

Straight Beam

Beam with straight profile and majority of beams in structures are straight beams.

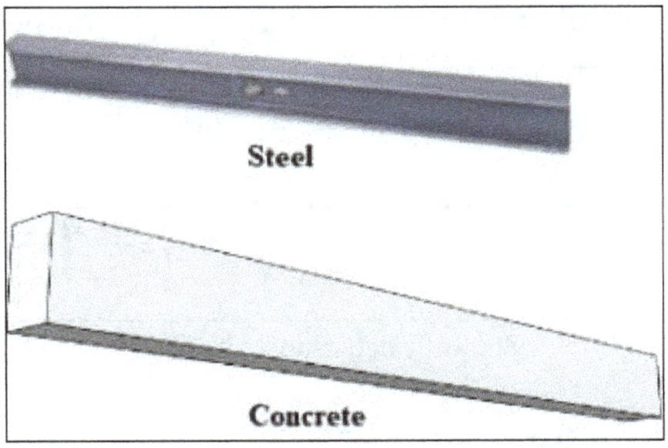

Straight beam.

Curved Beam

Beam with curved profile, such as in the case of circular buildings.

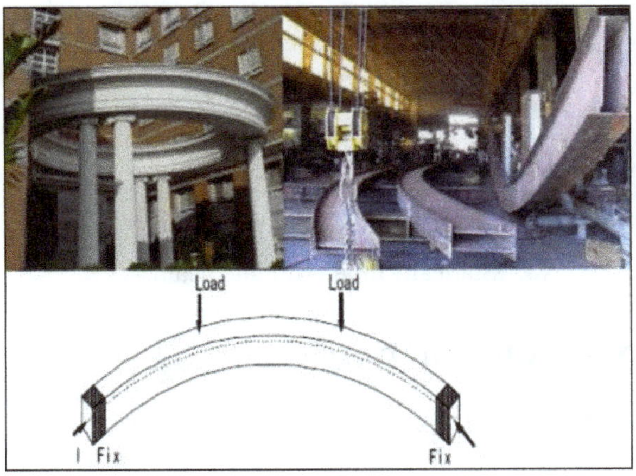

Curved beam

Tapered Beam

Beam with tapered cross-section.

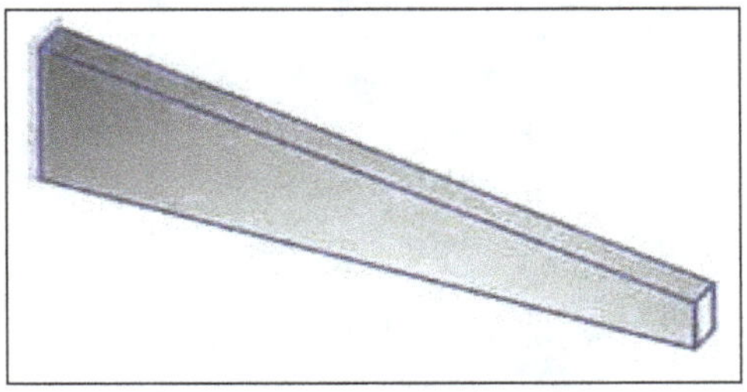

Tapered beam.

Based on Equilibrium Condition

Statically Determinate Beam

For a statically determinate beam, equilibrium conditions alone can be used to solve reactions, i.e. the numbers of unknown reactions are equal to the number of equations.

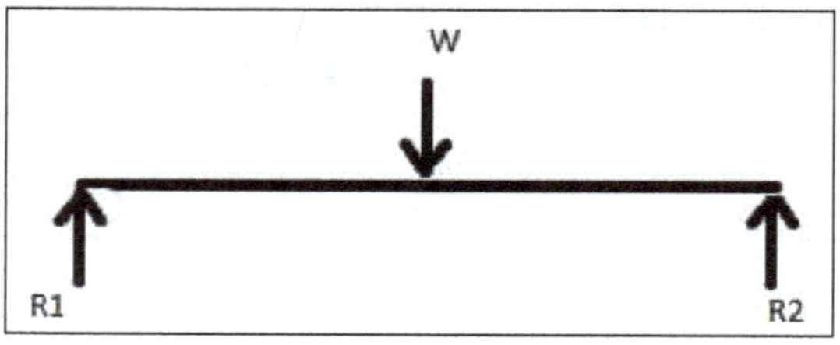

Statically determinate beam.

Statically Indeterminate Beam

For a statically indeterminate beam, equilibrium conditions are not enough to solve reactions. So, the analysis of this type of beam is more complicated than that of statically determinate beams.

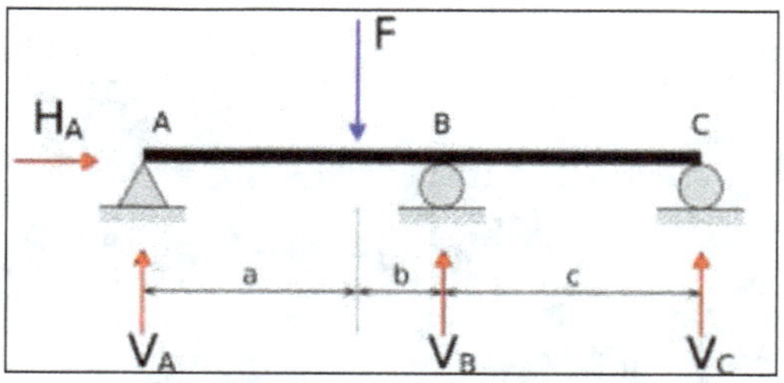

Statically indeterminate beam.

Based on Method of Construction

Cast In-situ Concrete Beam

This type of beam is constructed on project site. So, forms are initially fixed then fresh concrete is poured and allowed to be hardened. Then, loads will be imposed.

Cast in situ beam.

Precast Concrete Beam

This type of beam is manufactured in factories. So, the construction condition is more controllable compare with on-site construction. Consequently, the quality of concrete of the beam will be greater. Various cross-sectional shapes can be manufactures such as T- beam, Double T-beam, Inverted T-beam and many more.

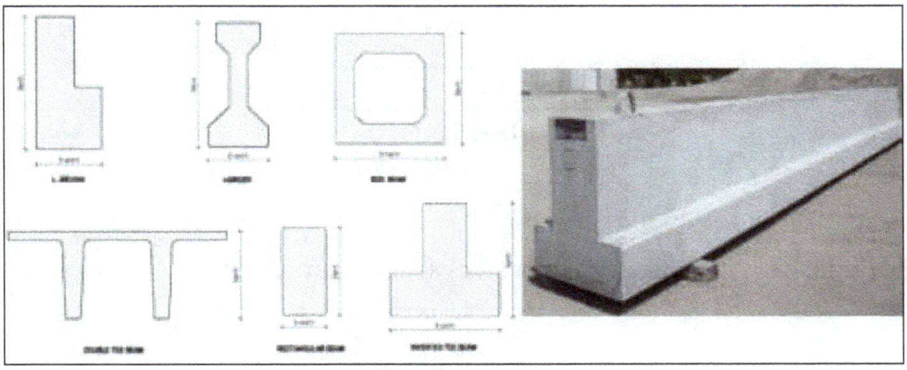

Precast concrete beam.

Prestressed Concrete Beam

This type of beam constructed by stressing strands prior to applying loads on the beam. Pre-tensioned Concrete beam and post-tensioned Concrete Beam are variations of pre-stressed concrete beam.

Prestressed concrete beam.

Others

Deep Beam

Beams that have large depths and its clear span to depth ratio is less than 4 according to ACI Code. Significant amount of the load is carried to the supports by a compression force combining the load and the reaction. Consequently, the strain distribution is no longer considered linear as in the case of conventional beams.

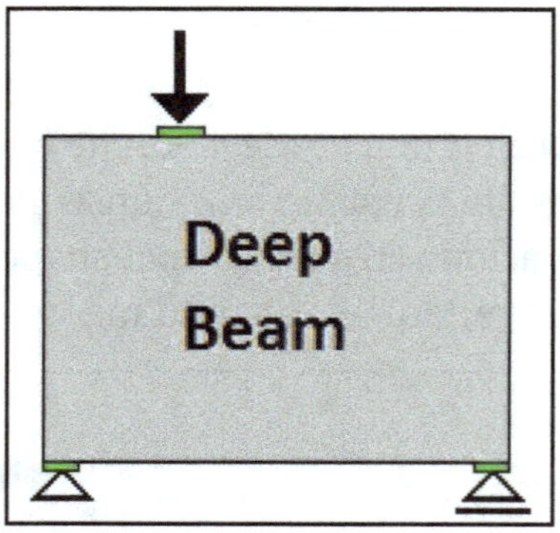

Deep beam.

Girder

Beams that take heavy loads, generally steel sections are used.

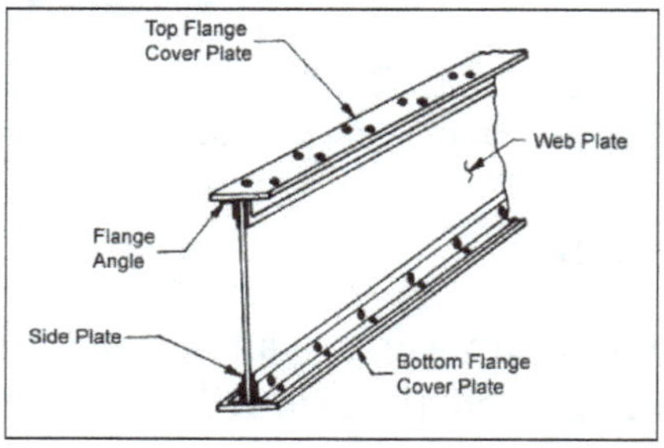

Girder.

Stress and Deflection in Beams

Many structures can be approximated as a straight beam or as a collection of straight beams. For this reason, the analysis of stresses and deflections in a beam is an important and useful topic.

Shear Force and Bending Moment

To find the shear force and bending moment over the length of a beam, first solve for the external reactions at the boundary conditions. For example, the cantilever beam below has an applied force shown in red, and the reactions are shown in blue at the fixed boundary condition.

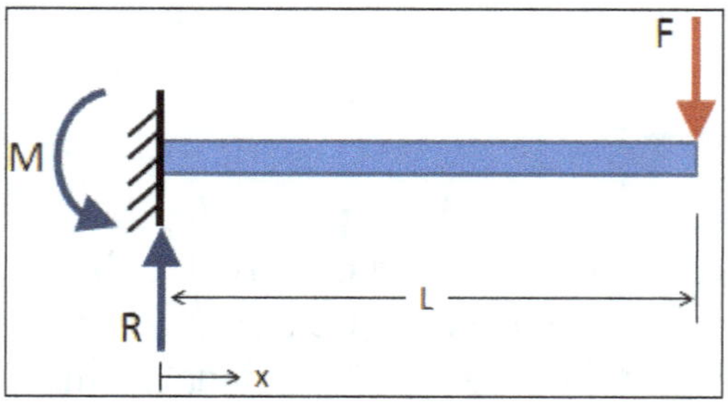

After the external reactions have been solved for, take section cuts along the length of the beam and solve for the reactions at each section cut. An example section cut is shown in the figure below.

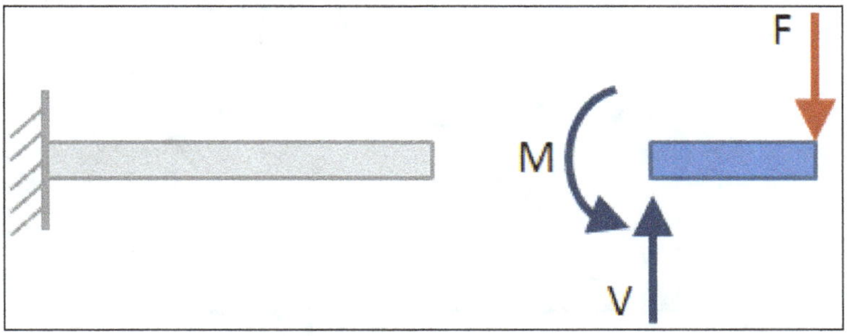

When the beam is cut at the section, either side of the beam can be considered when solving for the reactions. The side that is selected does not affect the results, so choose whichever side is easiest. In the figure above, the side of the beam to the right of the section cut was selected. The reactions at the section cut are shown with blue arrows.

Sign Convention

The signs of the shear and moment are important. The sign is determined after a section cut is taken and the reactions are solved for the portion of the beam to one side of the cut. The shear force at the section cut is considered positive if it causes clockwise rotation of the selected beam section, and it is considered negative if it causes counter-clockwise rotation. The bending moment at the section cut is considered positive if it compresses the top of the beam and elongates the bottom of the beam (i.e. if it makes the beam "smile").

Based on this sign convention, the shear force at the section cut in the figure above is positive since it causes clockwise rotation of the selected section. The moment is negative since it compresses the bottom of the beam and elongates the top (i.e. it makes the beam "frown").

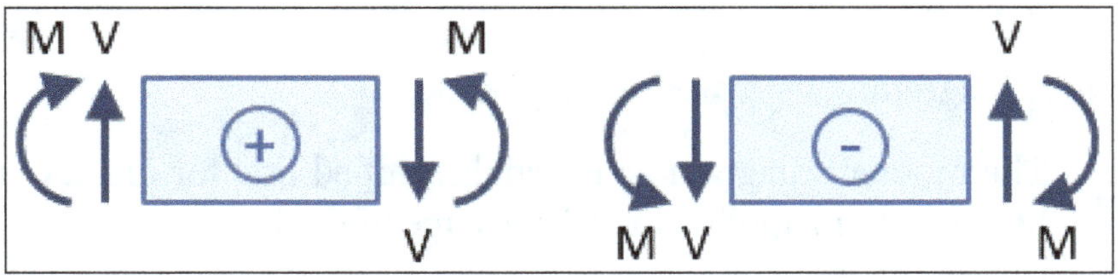

Shear and Moment Diagrams

The shear and bending moment throughout a beam are commonly expressed with diagrams. A shear diagram shows the shear along the length of the beam, and a moment diagram shows the bending moment along the length of the beam. These diagrams are typically shown stacked on top of one another, and the combination of these two diagrams is a shear-moment diagram. Shear-moment diagrams for some common end conditions and loading configurations are shown within the beam deflection tables. An example of a shear-moment diagram is shown in the following figure.

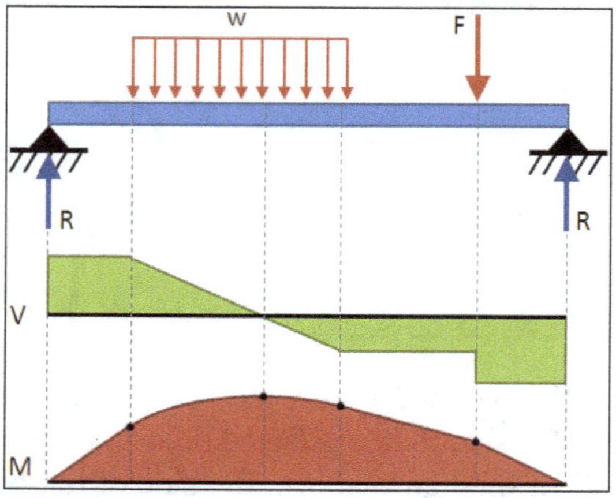

General rules for drawing shear-moment diagrams are given below:

- Shear Diagram:
    - Point loads cause a vertical jump in the shear diagram. The direction of the jump is the same as the sign of the point load.
    - Uniform distributed loads result in a straight, sloped line on the shear diagram. The slope of the line is equal to the value of the distributed load.
    - The shear diagram is horizontal for distances along the beam with no applied load.
    - The shear at any point along the beam is equal to the slope of the moment at that same point:

      $$V = \frac{dM}{dx}$$
- Moment Diagram:
    - The moment diagram is a straight, sloped line for distances along the beam with no applied load. The slope of the line is equal to the value of the shear.
    - Uniform distributed loads result in a parabolic curve on the moment diagram.
    - The maximum/minimum values of moment occur where the shear line crosses zero.

- The moment at any point along the beam is equal to the area under the shear diagram up to that point:

  $$M = \int V\,dx$$

Bending Stresses in Beams

The bending moment, M, along the length of the beam can be determined from the moment diagram. The bending moment at any location along the beam can then be used to calculate the bending stress over the beam's cross-section at that location. The bending moment varies over the height of the cross-section according to the flexure formula below:

$$\sigma_b = \frac{My}{I_c}$$

Where, M is the bending moment at the location of interest along the beam's length, I_c is the centroidal moment of inertia of the beam's cross-section, and y is the distance from the beam's neutral axis to the point of interest along the height of the cross-section. The negative sign indicates that a positive moment will result in a compressive stress above the neutral axis.

The bending stress is zero at the beam's neutral axis, which is coincident with the centroid of the beam's cross-section. The bending stress increases linearly away from the neutral axis until the maximum values at the extreme fibers at the top and bottom of the beam.

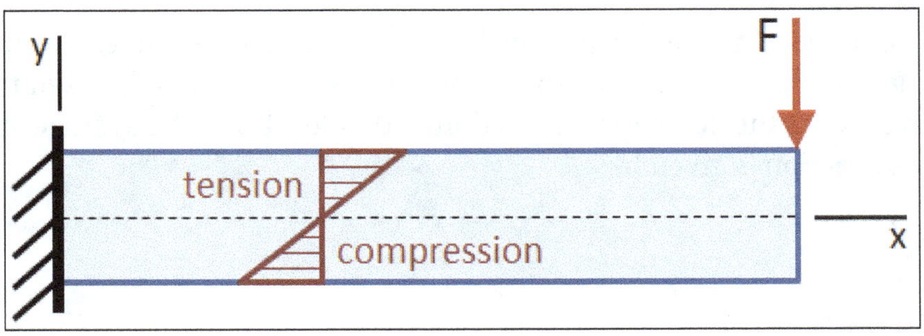

The maximum bending stress is given by:

$$\sigma_{b.max} = \frac{Mc}{I_c}$$

Where, c is the centroidal distance of the cross-section (the distance from the centroid to the extreme fiber).

If the beam is asymmetric about the neutral axis such that the distances from the neutral axis to the top and to the bottom of the beam are not equal, the maximum stress will occur at the farthest location from the neutral axis. In the figure below, the tensile stress at the top of the beam is larger than the compressive stress at the bottom.

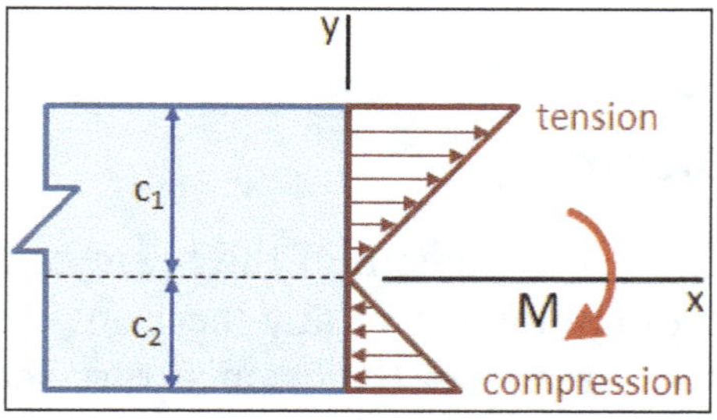

The section modulus of a cross-section combines the centroidal moment of inertia, I_c, and the centroidal distance, c:

$$S = \frac{I_c}{c}$$

The benefit of the section modulus is that it characterizes the bending resistance of a cross-section in a single term. The section modulus can be substituted into the flexure formula to calculate the maximum bending stress in a cross-section:

$$\sigma_{b.max} = \frac{M}{S}$$

Shear Stresses in Beams

The shear force, V, along the length of the beam can be determined from the shear diagram. The shear force at any location along the beam can then be used to calculate the shear stress over the beam's cross-section at that location. The average shear stress over the cross-section is given by:

$$\tau = \frac{V}{A}$$

The shear stress varies over the height of the cross-section, as shown in the figure below.

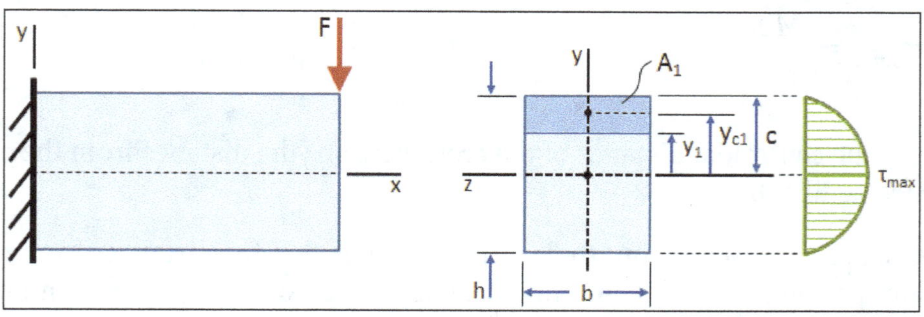

The shear stress is zero at the free surfaces (the top and bottom of the beam), and it is maximum at the centroid. The equation for shear stress at any point located a distance y_1 from the centroid of the cross-section is given by:

$$\tau = \frac{VQ}{I_c b}$$

Where, V is the shear force acting at the location of the cross-section, I_c is the centroidal moment of inertia of the cross-section, and b is the width of the cross-section. These terms are all constants. The Q term is the first moment of the area bounded by the point of interest and the extreme fiber of the cross-section:

$$Q = \int_{y1}^{c} y \, dA$$

Shear Stresses in Rectangular Sections

The distribution of shear stress along the height of a rectangular cross-section is shown in the figure below.

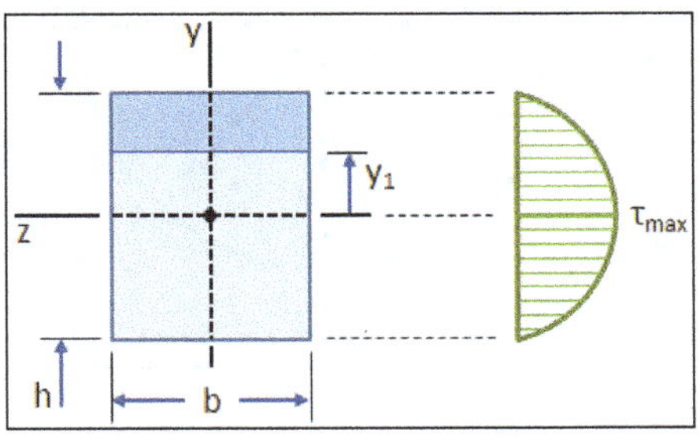

The first moment of area at any given point y_1 along the height of the cross-section is calculated by:

$$Q = \frac{b}{2}\left(\frac{h^2}{4} - y_1^2\right)$$

The maximum value of Q occurs at the neutral axis of the beam (where $y_1 = 0$):

$$Q_{max} = \frac{bh^2}{8}$$

The shear stress at any given point y1 along the height of the cross-section is calculated by:

$$\tau = \frac{VQ}{I_c b} = \frac{V}{2I_c}\left(\frac{h^2}{4} - y_1^2\right)$$

Where, $I_c = b{\cdot}h^3 / 12$ is the centroidal moment of inertia of the cross-section. The maximum shear stress occurs at the neutral axis of the beam and is calculated by:

$$\tau_{max} = \frac{3V}{2A}$$

Where, A = b·h is the area of the cross-section.

Note that the maximum shear stress in the cross-section is 50% higher than the average stress V/A.

Shear Stresses in Circular Sections

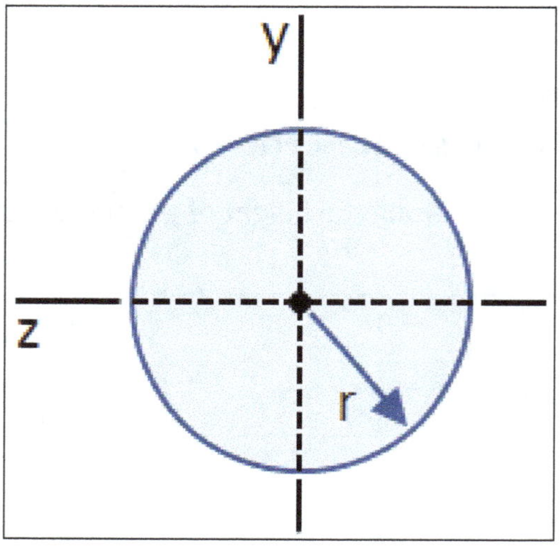

The equations for shear stress in a beam were derived using the assumption that the shear stress along the width of the beam is constant. This assumption is valid at the centroid of a circular cross-section, although it is not valid anywhere else. Therefore, while the distribution of shear stress along the height of the cross-section cannot be readily determined, the maximum shear stress in the section (occuring at the centroid) can still be calculated. The maximum value of first moment, Q, occurring at the centroid, is given by:

$$Q_{max} = \frac{2r^3}{3}$$

The maximum shear stress is then calculated by:

$$\tau_{max} = \frac{VQ_{max}}{I_c b} = \frac{4V}{3A}$$

Where, $b = 2r$ is the diameter (width) of the cross-section, $I_c = \pi r^4 / 4$ is the centroidal moment of inertia, and $A = \pi r^2$ is the area of the cross-section.

Shear Stresses in Circular Tube Sections

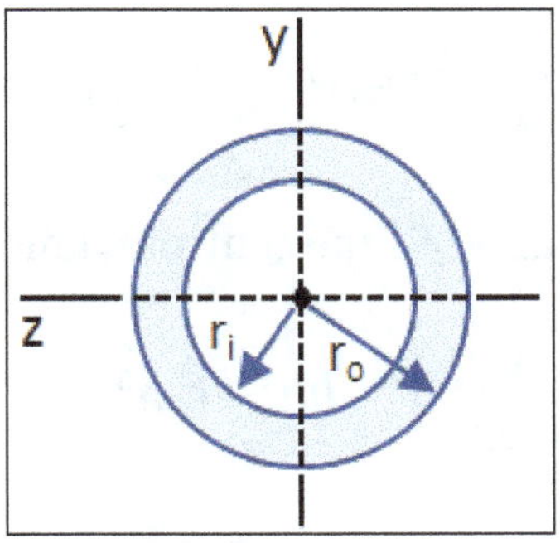

The maximum value of first moment, Q, occurring at the centroid, is given by:

$$Q_{max} = \frac{2}{3}(r_o^3 - r_i^3)$$

The maximum shear stress is then calculated by:

$$\tau_{max} = \frac{VQ_{max}}{I_c b} = \frac{4V}{3A}\left(\frac{r_o^2 + r_o r_i - r_i^2}{r_o^2 + r_i^2}\right)$$

Where, $b = 2\,(r_o - r_i)$ is the effective width of the cross-section, $I_c = \pi\,(r_o^4 - r_i^4)/4$ is the centroidal moment of inertia, and $A = \pi\,(r_o^2 - r_i^2)$ is the area of the cross-section.

Shear Stresses in I-Beams

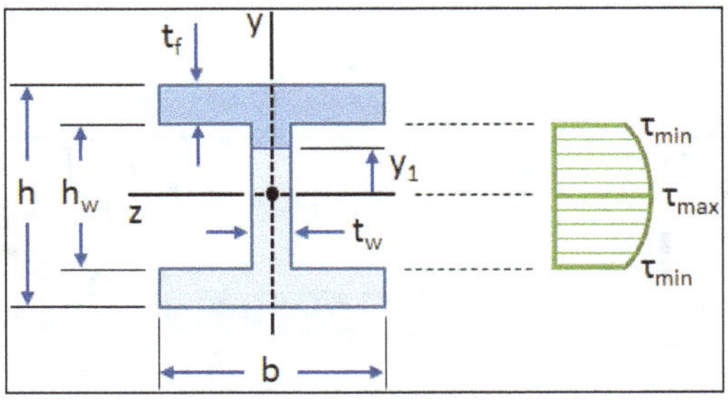

The equations for shear stress in a beam were derived using the assumption that the shear stress along the width of the beam is constant. This assumption is valid over the web of an I-Beam, but it is invalid for the flanges (specifically where the web intersects the flanges). However, the web of an I-Beam takes the vast majority of the shear force (approximately 90% - 98%, according to Gere), and so it can be conservatively assumed that the web carries all of the shear force.

The first moment of the area of the web of an I-Beam is given by:

$$Q = \frac{b}{8}\left(h^2 - h_w^2\right) + \frac{t_w}{8}\left(h_w^2 - 4y_1^2\right)$$

The shear stress along the web of the I-Beam is given by:

$$\tau = \frac{VQ}{I_c t_w} = \frac{V}{8 I_c t_w}\left[b\left(h^2 - h_w^2\right) + t_w\left(h_w^2 - 4y_1^2\right)\right]$$

Where, t_w is the web thickness and I_c is the centroidal moment of inertia of the I-Beam:

$$I_c = \frac{bh^3}{12} - \frac{(b - t_w)h_w^3}{12} = \frac{1}{12}\left(bh^3 - bh_w^3 + t_w h_w^3\right)$$

The maximum value of shear stress occurs at the neutral axis ($y_1 = 0$), and the minimum value of shear stress in the web occurs at the outer fibers of the web where it intersects the flanges $y_1 = \pm h_w/2$:

$$\tau_{max} = \frac{V}{8 I_c t_w}\left(bh^2 - bh_w^2 + t_w h_w^2\right)$$

$$\tau_{max} = \frac{V}{8 I_c t_w}\left(h^2 - h_w^2\right)$$

Beam Deflection Tables

The tables below give equations for the deflection, slope, shear, and moment along straight beams for different end conditions and loadings.

Cantilever Beams

Cantilever: End Load

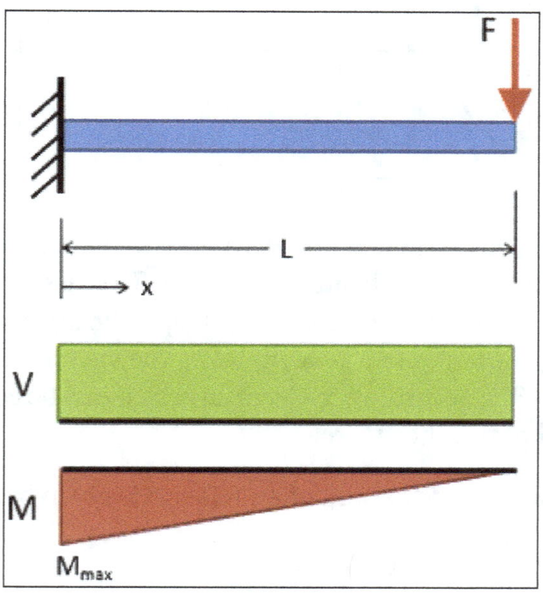

Deflection:

$$\delta = -\frac{Fx^2}{6EI}(3L - x)$$

$$\delta_{max} = \frac{FL^3}{3EI} \ @x = L$$

Slope:

$$\theta = -\frac{Fx}{2EI}(2L - x)$$

$$\theta_{max} = \frac{FL^2}{2EI} \ @x = L$$

Shear:

$$V = +F$$

Moment:

$$M = -F(\mathrm{L} - x)$$

$$M_{max} = -FL \ @x = 0$$

Cantilever: Intermediate Load

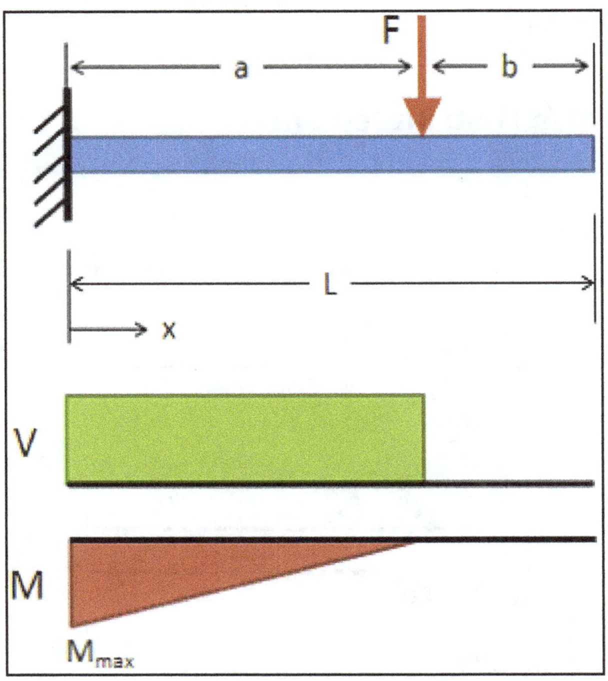

Deflection:

$$\delta = -\frac{Fx^2}{6EI}(3a-x)\ (0 \le x \le a)$$

$$\delta = -\frac{Fa^2}{6EI}(3x-a)\ (a \le x \le L)$$

$$\delta_{max} = \frac{Fa^2}{6EI}(3L-a)\ @x=L$$

Slope:

$$\theta = -\frac{Fx}{2EI}(2a-x)\ (0 \le x \le a)$$

$$\theta = -\frac{Fa^2}{2EI}\ (a \le x \le L)$$

Shear:

$$V = +F\ (0 \le x \le a)$$

$$V = 0\ (a \le x \le L)$$

Moment:

$$M = -F(a-x)\ (0 \le x \le a)$$

$$M = 0\ (a \le x \le L)$$

Cantilever: Uniform Distributed Load

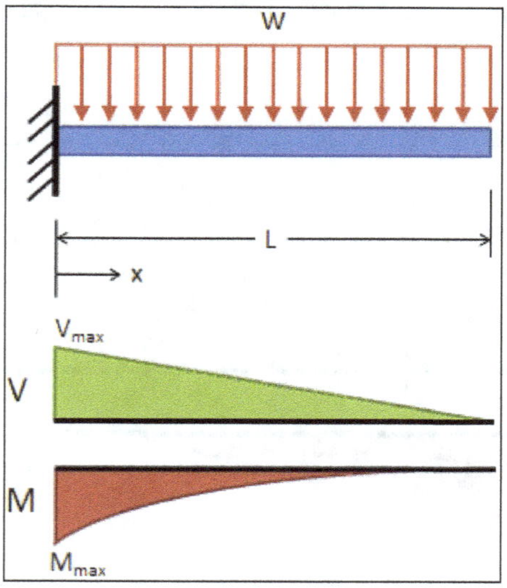

Deflection:

$$\delta = -\frac{wx^2}{24EI}\left(6L^2 - 4Lx + x^2\right)$$

$$\delta_{max} = \frac{wL^4}{8EI} \ @x = L$$

Slope:

$$\theta = -\frac{wx}{6EI}\left(3L^2 - 3Lx + x^2\right)$$

$$\theta_{max} = \frac{wL^3}{6EI} \ @x = L$$

Shear:

$$V = +w\left(L - x\right)$$

$$V_{max} = +wL \ @x = L$$

Moment:

$$M = -w\left(L - x\right)^2 / 2$$

$$M_{max} = -wL^2 / 2 \ @x = 0$$

Cantilever: Triangular Distributed Load

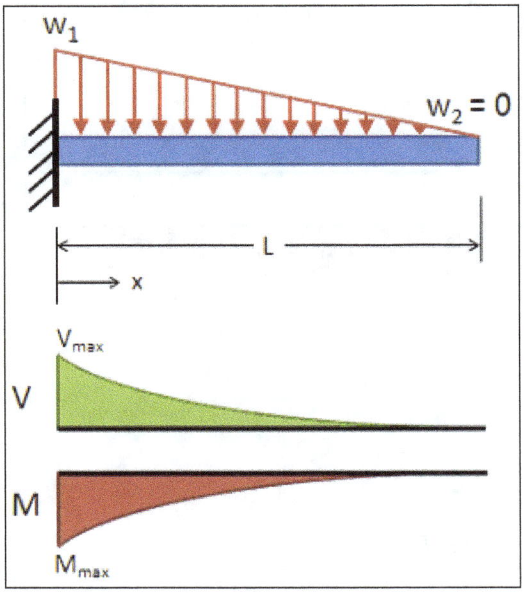

Deflection:

$$\delta = -\frac{w_1 x^2}{120LEI}\left(10L^3 - 10L^2x + 5Lx^2 - x^3\right)$$

$$\delta_{max} = \frac{w_1 L^4}{30EI} \ @x = L$$

Slope:

$$\theta = -\frac{w_1 L}{24LEI}\left(4L^3 - 6L^2x + 4Lx^2 - x^3\right)$$

$$\theta_{max} = \frac{w_1 L^3}{24EI} \ @x = L$$

Shear:

$$V_{max} = +w_1 L/2 \ @x = L$$

Moment:

$$M_{max} = -w_1 L^2/6 \ @x = 0$$

Cantilever: End Moment

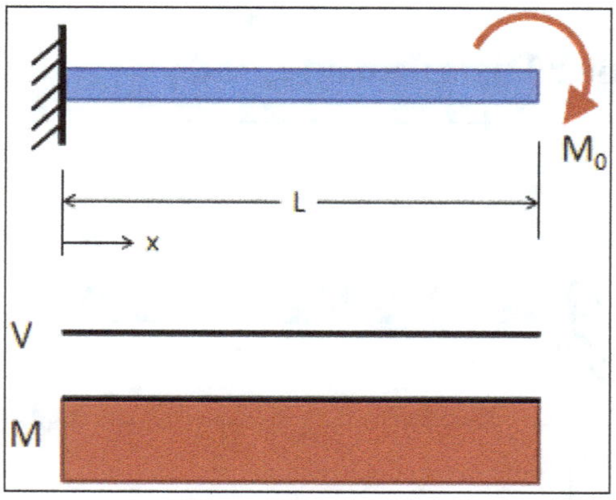

Deflection:

$$\delta = -\frac{M_0 x^2}{2EI}$$

$$\delta_{max} = \frac{M_0 L^2}{EI} \ @x = L$$

Slope:

$$\theta = -\frac{M_o x}{EI}$$

$$\theta_{max} = \frac{M_o L}{EI} \quad @x = L$$

Shear:

$$V = 0$$

Moment:

$$M = -M_o$$

Simply Supported Beams

Simply Supported: Intermediate Load

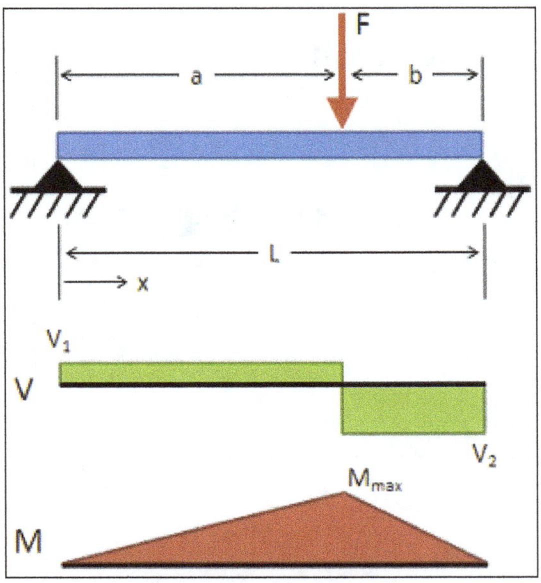

Deflection:

$$\delta = -\frac{Fbx}{6LEI}\left(L^2 - b^2 - x^2\right) \quad (0 \le x \le a)$$

For $a \ge b$:

$$\delta_{max} = \frac{Fb\left(L^2 - b^2\right)^{3/2}}{9\sqrt{3}LEI} \quad @x = \sqrt{\frac{L^2 - b^2}{3}}$$

Slope:

$$\theta = -\frac{Fb}{6LEI}\left(L^2 - b^2 - 3x^2\right)\ (0 \le x \le a)$$

$$\theta_1 = -\frac{Fab(L+b)}{6LEI}\ @x=0$$

$$\theta_2 = \frac{Fab(L+a)}{6LEI}\ @x=L$$

Shear:

$$V_1 = +Fb/L\ \ (0 \le x \le a)$$

$$V_2 = -Fa/L\ \ (a \le x \le L)$$

Moment:

$$M_{max} = +Fab/L\ \ @x=a$$

Simply Supported: Center Load

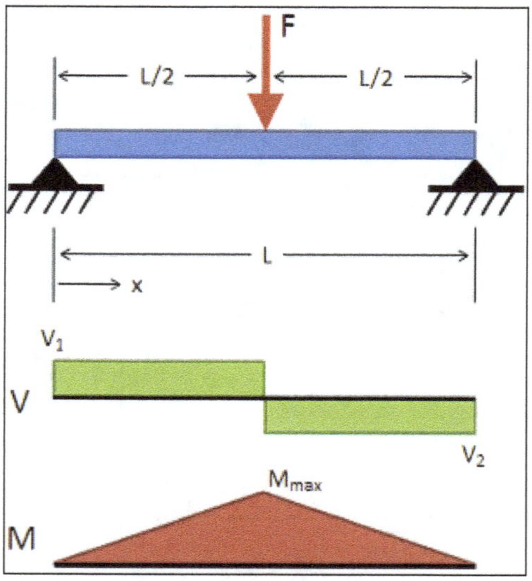

Deflection:

$$\delta = -\frac{Fx}{48EI}\left(3L^2 - 4x^2\right)\ \ (0 \le x \le L/2)$$

$$\delta_{max} = \frac{FL^3}{48EI}\ \ @x=L/2$$

Slope:

$$\theta = -\frac{F}{16EI}\left(L^2 - 4x^2\right) \ \left(0 \le x \le L/2\right)$$

$$\theta_1 = -\frac{FL^2}{16EI} \quad @x = 0$$

$$\theta_2 = +\frac{FL^2}{16EI} \quad @x = L$$

Shear:

$$V_1 = +F/2 \ \left(0 \le x \le L/2\right)$$

$$V_2 = -F/2 \ \left(L/2 \le x \le L\right)$$

Moment:

$$M_{\max} = FL/4 \quad @x = L/2$$

Simply Supported: 2 Loads at Equal Distances from Supports

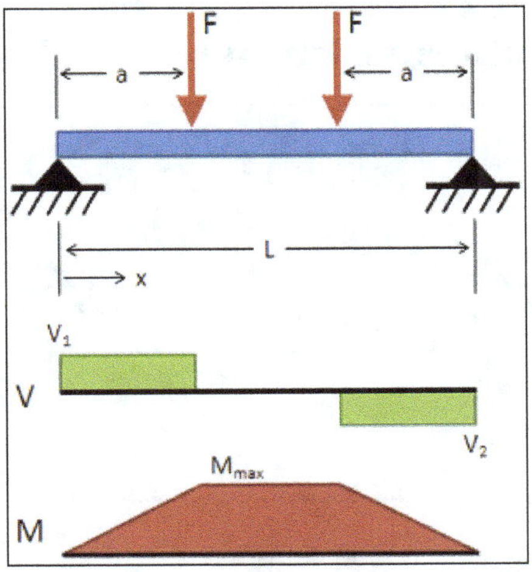

Deflection:

$$\delta = -\frac{Fx}{6EI}\left(3aL - 3a^2 - x^2\right) \ \left(0 \le x \le a\right)$$

$$\delta = -\frac{Fa}{6EI}\left(3Lx - 3x^2 - a^2\right) \ \left(a \le x \le L - a\right)$$

$$\delta_{max} = \frac{Fa}{24EI}\left(3L^2 - 4a^2\right) \quad @x = L/2$$

Slope:

$$\theta = -\frac{F}{2EI}\left(aL - a^2 - x^2\right)\ (0 \le x \le a)$$

$$\theta = -\frac{Fa}{2EI}(L - 2x)\ \ (a \le x \le L - a)$$

$$\theta_1 = -\frac{Fa(L-a)}{2EI}\ \ @x=0$$

$$\theta_2 = +\frac{Fa(L-a)}{2EI}\ \ @x=L$$

Shear:

$$V_1 = +F\ \ (0 \le x \le a)$$

$$V_2 = -F\ \ (L - a \le x \le L)$$

Moment:

$$M_{\max} = Fa\ \ (a \le x \le L - a)$$

Simply Supported: Uniform Distributed Load

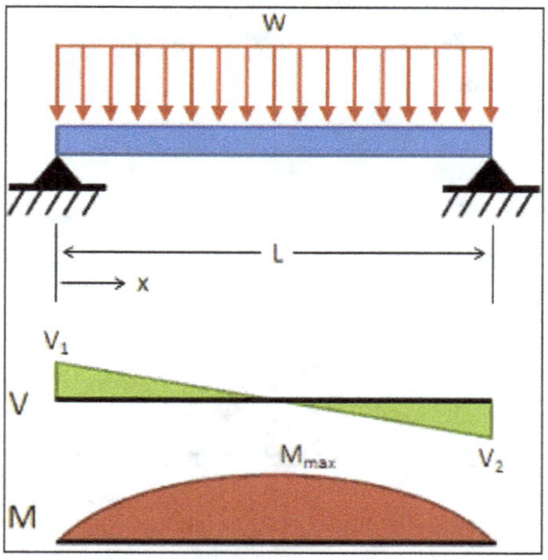

Deflection:

$$\delta = -\frac{wx}{24EI}\left(L^3 - 2Lx^2 - x^3\right)$$

$$\delta_{max} = \frac{5wL^4}{384EI}\ \ @x=L/2$$

Slope:

$$\theta = -\frac{w}{24EI}\left(L^3 - 6Lx^2 - 4x^3\right)$$

$$\theta_1 = -\frac{wL^3}{24EI} \quad @x=0$$

$$\theta_2 = +\frac{wL^3}{24EI} \quad @x=L$$

Shear:

$$V = w\left(L/2 - x\right)$$

$$V_1 = +wL/2 \quad @x=0$$

$$V_2 = -wL/2 \quad @x=L$$

Moment:

$$M_{\max} = wL^2/8 \quad @x=L/2$$

Simply Supported: Moment at each Support

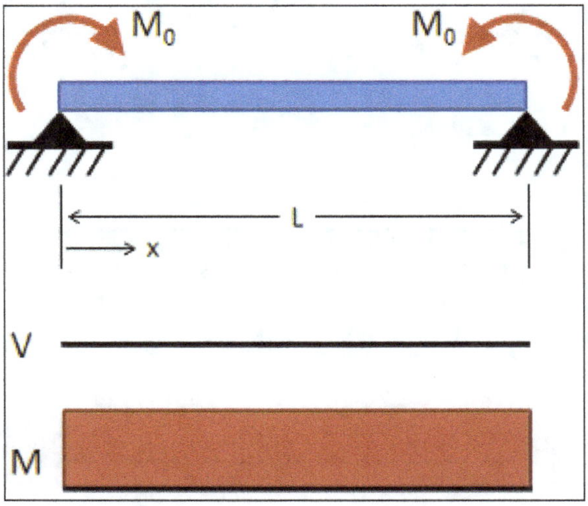

Deflection:

$$\delta = -\frac{M_0 x}{2EI}\left(L - x\right)$$

$$\delta_{max} = \frac{M_0 L^2}{8EI} \quad @x=L/2$$

Slope:

$$\theta = -\frac{M_o}{2EI}(L - 2x)$$

$$\theta_1 = -\frac{M_o L}{2EI} \quad @x=0$$

$$\theta_2 = +\frac{M_o L}{2EI} \quad @x=L$$

Shear:

$$V = 0$$

Moment:

$$M = M_o$$

Simply Supported: Moment at One Support

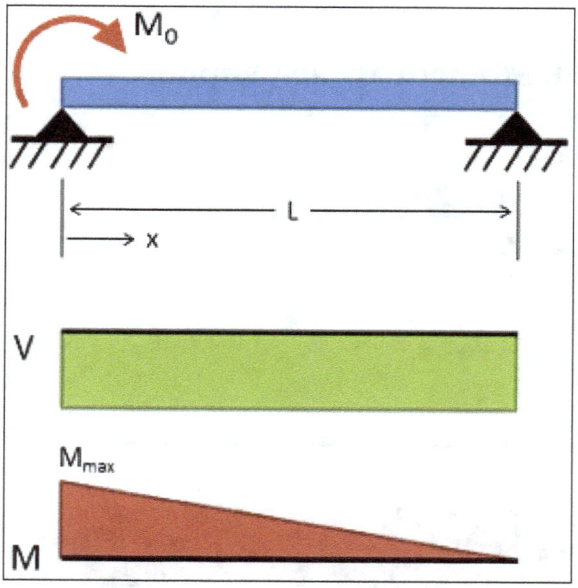

Deflection:

$$\delta = -\frac{M_o x}{6LEI}\left(2L^2 - 3Lx + x^2\right)$$

$$\delta_{max} = \frac{M_o L^2}{9\sqrt{3}EI} \quad @x=L\left(1-\sqrt{3}/3\right)$$

Slope:

$$\theta = -\frac{M_o}{6LEI}\left(2L^2 - 6Lx + 3x^2\right)$$

$$\theta_1 = -\frac{M_o L}{3EI} \quad @x=0$$

$$\theta_2 = +\frac{M_o L}{6EI} \quad @x=L$$

Shear:

$$V = -M_o / L$$

Moment:

$$M_{max} = M_o \quad @x=0$$

Simply Supported: Center Moment

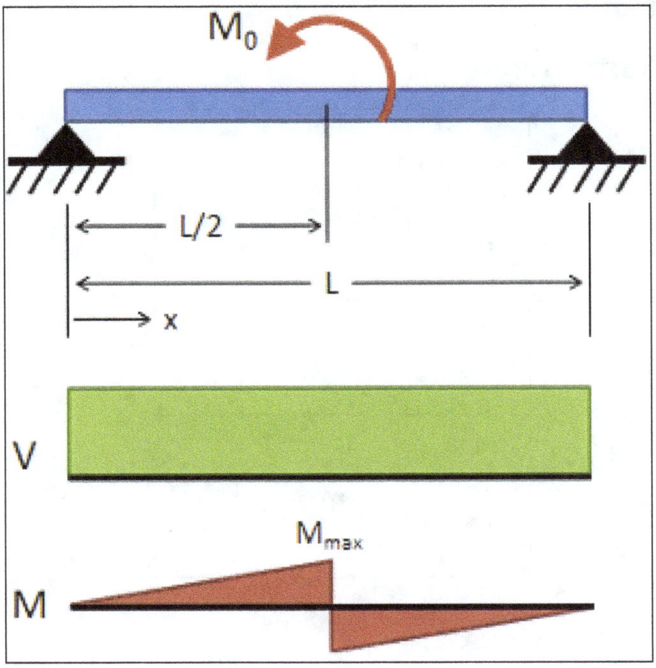

Deflection:

$$\delta = -\frac{M_o x}{24LEI}\left(L^2 - 4x^2\right) \quad \left(0 \le x \le L/2\right)$$

Slope:

$$\theta = -\frac{M_o}{24LEI}\left(L^2 - 12x^2\right)\ \left(0 \le x \le L/2\right)$$

$$\theta_1 = -\frac{M_o L}{24EI}\ \ @x=0$$

$$\theta_2 = -\frac{M_o L}{24EI}\ \ @x=L$$

Shear:

$$V = +M_o / L$$

Moment:

$$\mathrm{M} = \mathrm{M}_o\, x/L\ \left(0 \le x \le L/2\right)$$

$$M_{\max} = M_o / 2\ \ @x=L/2$$

Fixed-Fixed Beams

Fixed-Fixed: Center Load

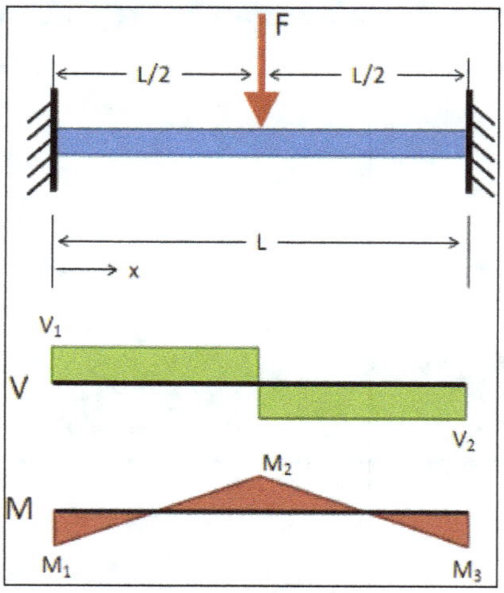

Deflection:

$$\delta = -\frac{Fx^2}{48EI}(3L - 4x)\ \left(0 \le x \le L/2\right)$$

$$\delta_{max} = -\frac{FL^3}{192EI}\ \ @x=L/2$$

Shear:

$$V_1 = +F/2 \ (0 \le x \le L/2)$$
$$V_2 = -F/2 \ (L/2 \le x \le L)$$

Moment:

$$M = F(4x - L)/8 \ (0 \le x \le L/2)$$
$$M_1 = M_3 = -FL/8 \ \ @x = 0 \text{ and } x = L$$
$$M_2 = +FL/8 \ \ @x = L/2$$

Fixed-Fixed: Uniform Distributed Load

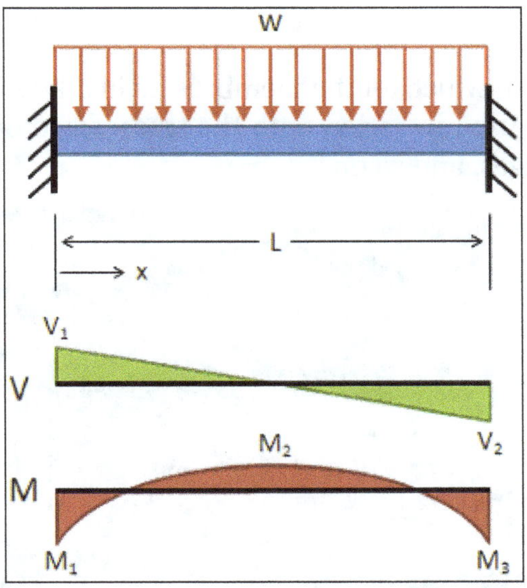

Deflection:

$$\delta = -\frac{wx^2}{24EI}(L - x)^2$$
$$\delta_{max} = -\frac{wL^4}{384EI} \ \ @x = L/2$$

Shear:

$$V = w(L/2 - x)$$
$$V_1 = +wL/2 \ \ @x = 0$$
$$V_2 = -wL/2 \ \ @x = L$$

Moment:

$$M = w\left(6Lx - 6x^2 - L^2\right)/12$$

$$M_1 = M_3 = -wL^2/12 \quad @x = 0 \text{ and } x = L$$

$$M_2 = -wL^2/24 \quad @x = L/2$$

Types of Truss Structures

Simple Truss

This is a single triangle such as might be found in a framed roof consisting of rafters and a ceiling joist.

Planar Truss

A planar truss is a truss in which all the members lie in a two-dimensional plane. This type of truss is typically used in series, with the trusses laid out in a parallel arrangement to form roofs, bridges, and so on.

Space Frame Truss

In contrast to a planar truss which lies in a two-dimensional plane, a space frame truss is a three-dimensional framework of connected triangles.

Truss Forms

There are a wide range of truss forms that can be created, varying in materials, overall geometry and span.

Pratt Truss

Also known as an 'N' truss, this form is often used in long-span buildings, with spans ranging from 20-100 m, where uplift loads may be predominant, such as in aircraft hangers. A Pratt truss uses vertical members for compression and horizontal members for tension. The configuration of the members means that longer diagonal members are only in tension for gravity load effects which allows them to be used more efficiently.

Warren Truss

A Warren truss has fewer members than a Pratt truss and has diagonal members which are alternatively in tension and compression. The truss members form a series of equilateral triangles, alternating up and down.

North Light Truss

This form of truss is usually used for short spans in industrial buildings, and is so called because it allows maximum benefit to be gained from natural lighting by the use of glazing on the steeper north-facing pitch (sometimes referred to as a sawtooth roof). It is common, on the steeper sloping portion of the truss, to have a second truss running perpendicular to the plane of the north light truss, providing large column-free space.

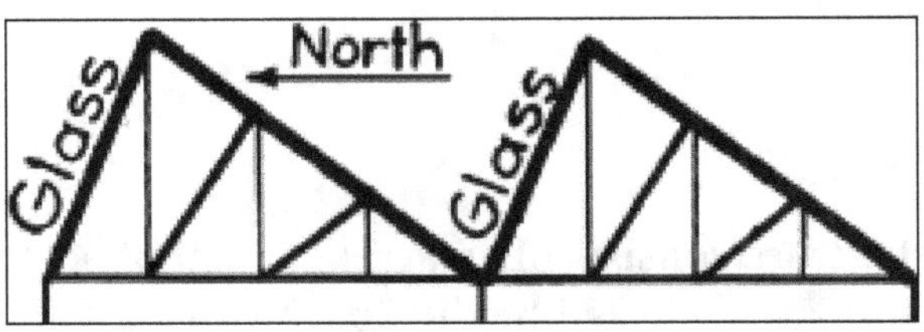

King Post Truss

Typically made from timber, and spanning up to 8m, king post trusses are commonly used in the construction of domestic roofs. They take the form of a simple triangle, with a vertical member between the apex and the bottom chord.

Queen Post Truss

Similar to the king post truss, but with diagonal members between the centre of the bottom chord and each of the inclined top chords, queen post trusses can span 10m.

Flat Truss

The top and bottom chords are parallel, allowing the construction of floors or flat roofs.

Other Types of Truss

Other variations include:

- Howe truss
- Scissor roof truss
- Hip truss
- Bowstring truss
- Fan truss
- Fink Truss

Shells

The typical modem structure consists of an arrangement of beams and columns. The loads in such a structure are collected by the flooring or roof system and distributed into the beams. These beams then transmit the vertical load from the point of entry to the ends of the beam. At this point the load is transferred into the columns. The load is then transmitted along the column length to the foundation system, where it is distributed into the ground surface. In the analysis of the beams and columns, these elements are treated as lines with a certain cross-section. The stresses acting on the cross-section can then be calculated based upon a number of simplifying assumptions. These structural elements fall into the category of the linear, or one-dimensional, members. At the other end of the spectrum is the three dimensional elastic continua. Without any geometric limitations, no simplifying assumptions can be reasonably made.

In between these two extremes lies the class of structures known as two-dimensional surface structures. Surface structures are, like linear members, capable of transmitting loads from one point to another. Whereas the load path in a linear member is along the line, the load path in a surface structure is along the surface. Included in this class of surface structures are plates and shells. Both are defined as having one dimension, namely the thickness, significantly smaller than the other two. The definition of a plate requires that its entire surface lie in a plane. On the other hand, a shell is defined as a curved surface structure.

Concrete Thin Shell Structure

Concrete thin shell structure is a three-dimensional spatial structure that constructed from one or more curved slabs or folded plates. The thicknesses of curved slab and folded plates are small compared to their other dimensions.

The outstanding features of concrete thin shells are their three-dimensional load-carrying behavior which is governed by number of factors. For example, nature of the applied load, the manner in which thin shells are supported, and the geometry of thin shell forms.

Types and Forms of Shell Structure

Folded Plates

It is a type of concrete shell structure that made up of thin flat slabs. These slabs are jointed along their edges to construct three-dimensional structure.

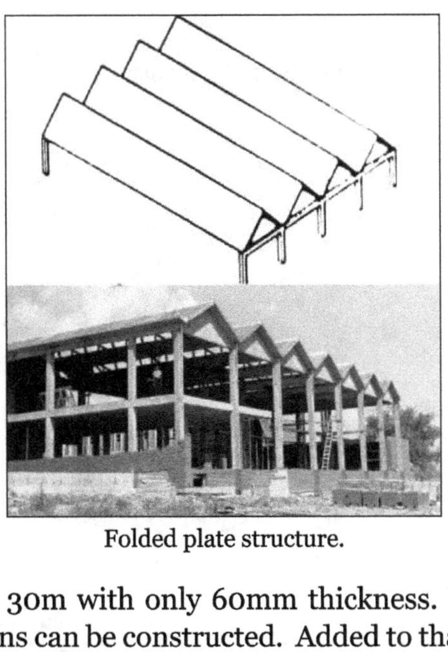

Folded plate structure.

Folded plate spans up to 30m with only 60mm thickness. Nonetheless, folded plate structure with greater spans can be constructed. Added to that, it is aesthetically pleasing, reduce material and construction cost.

However, difficult shuttering, requirement for extensive labor work and supervision, high cost, and difficulty in rising roofs are drawbacks of this type of plates.

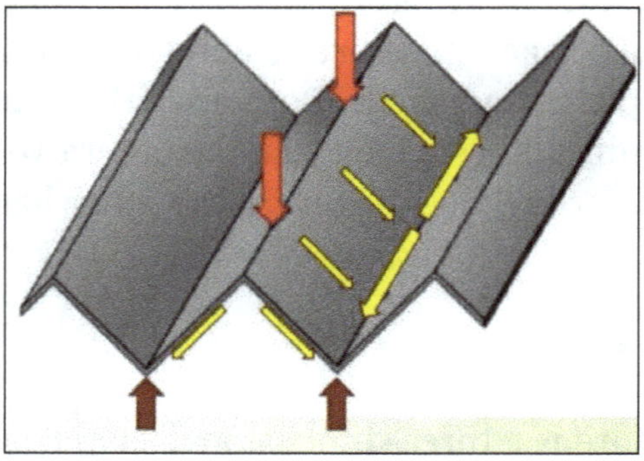

Distribution of loads in folded plate structure.

There are three types of folded plates which include:

1. Prismatic folded plates: It composites of planar thin slabs joining along their longitudinal edges to form a beam-like structure spanning between supports.

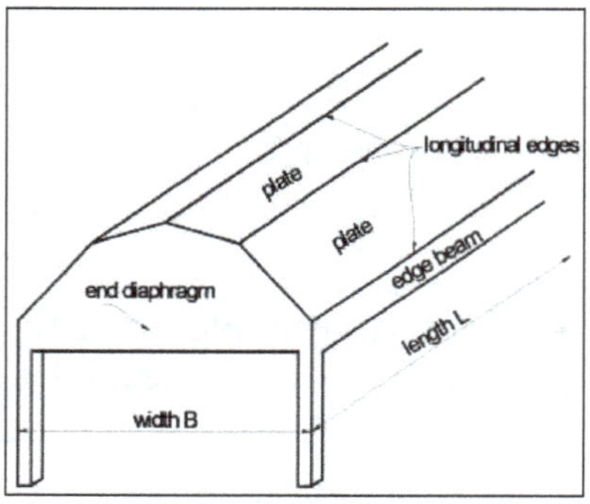

Prismatic folded plate.

2. Non-prismatic folded plates: It composites of planar thin slabs joining along their longitudinal edges to form a beam-like structure spanning between supports.

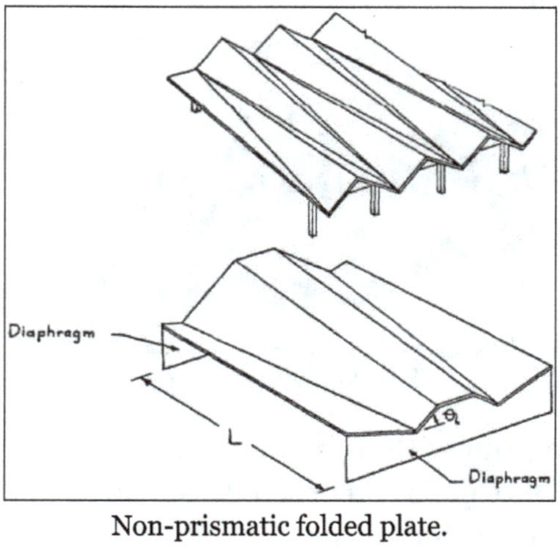

Non-prismatic folded plate.

3. Faceted folded plates: It consists of triangular or polygonal planar thin slabs joining along their edges to form thin plate structure.

Cylindrical Barrel Vaults

Basically, barrel vaults are deep concrete beams with considerably thin web member that can be designed using conventional reinforced concrete design.

Moreover, this type of concrete thin shell composes of cylinder, frame or ties at the ends including columns, and side elements which include a cylindrical element; a folded plate element; columns; or combination thereof.

Cylindrical barrel vault.

Furthermore, barrel vaults assumed to be very efficient due to the use the arch form to reduce stresses and thicknesses in the transverse direction. Added to that, it spans up to 45.72m that is why barrel vaults are the most useful shell structure.

Finally, a barrel shell carries load longitudinally as a beam and transversely as an arch. The arch, however, is supported by internal shears, and so may be calculated.

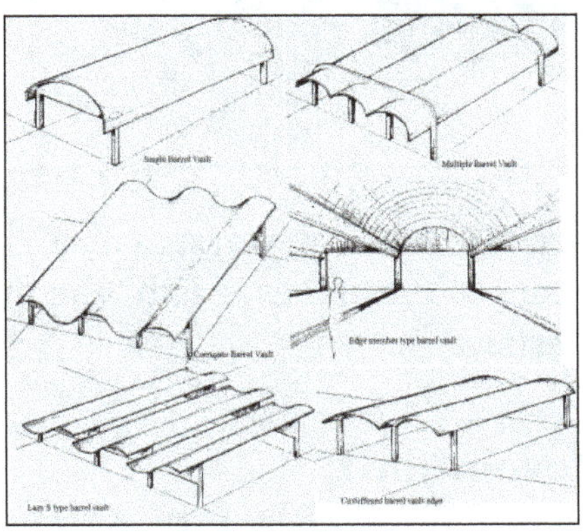

Different types of cylindrical barrel vault structure.

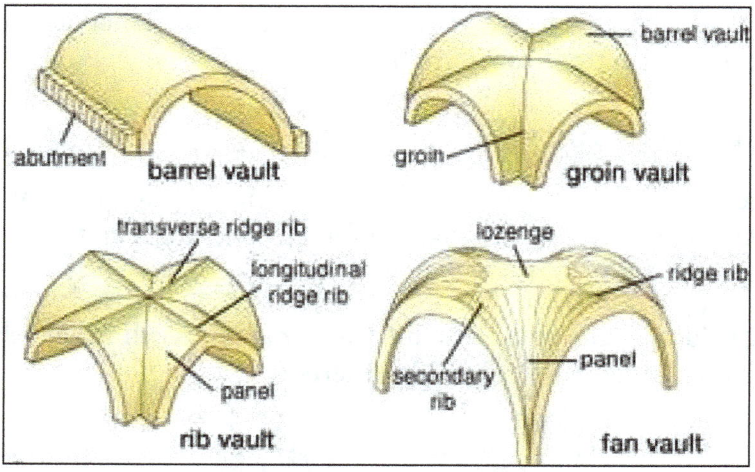

Different configuration and elements of barrel vault structure.

Short Shell

It is a cylindrical shell that has a considerably large radius in comparison with its length. Short shell consists of shell element spans between arches, arch structure, and edge beams provided at the lowest point of the shell. In small structure, it is permissible to omit edge beams in small structure.

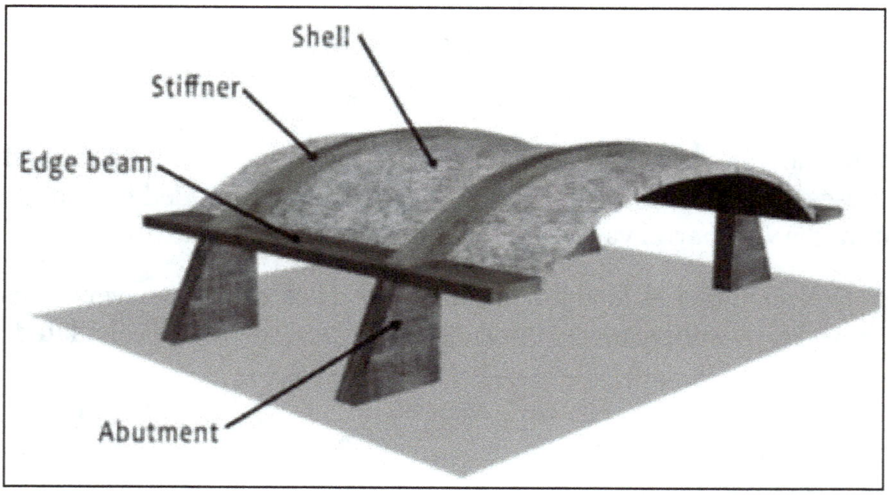

Moreover, the short shell carries load in two ways; as an arch carrying load to the lower elements and as a curved beam to the arches. Lastly, the thickness of the shell can be quite thin due to these properties.

Domes (Surfaces of Revolution)

They are structures that cover a more or less square or circular area. Domes are hemisphere in shape and used as a roof structure. Support element of domes includes columns, circular or regular polygon shaped wall.

Dome structure.

Moreover, they are membrane structures; the internal stresses are tension and compression and are statically determinate if the proper edge conditions are fulfilled.

Furthermore, in a dome of uniform thickness, under its own weight, the ring stresses are compression until the angle to the vertical is about 57 degrees.

Lastly, if the dome is less than a full hemisphere, a ring is required at the base of the dome to contain the forces.

Domes used in the construction of shopping center and sport center.

Folded Plate Domes

It is a type of thin concrete shell structure that consists of plane slabs and plates. Various configurations of folded plate domes are available and being constructed.

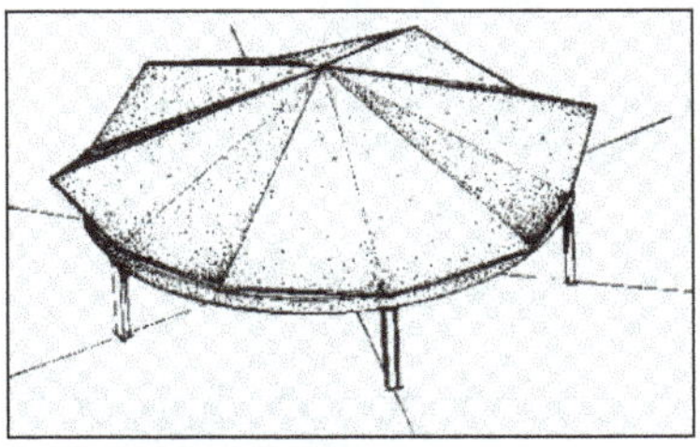

Folded plate domes.

Domes may be constructed with small angles between the plates or with large angles between plates and the structural action may be considerably different for each type.

Folded plate dome surfaces are easier to construct since they are flat. Nonetheless, for slab spans over 16 ft, the shell wall is thicker than a curved surface because bending must be considered.

The sound waves in areas covered with folded plate domes are not converge. Therefore, it is more desirable to the curved dome for use in an auditorium.

Finally, the structural design of folded plate domes follows that of folded plate barrels.

Folded plate dome structure.

Translation Shells

A translation shell is a dome set on four arches. The shape is different from a spherical dome and is generated by a vertical circle moving on another circle. All vertical slices have the same radius. It is easier to form than a spherical dome.

The stresses in a translation shell are much like a dome at the top, but at the level of the arches, tension forces are offset by compression in the arch. However, there are high tension forces in the corner.

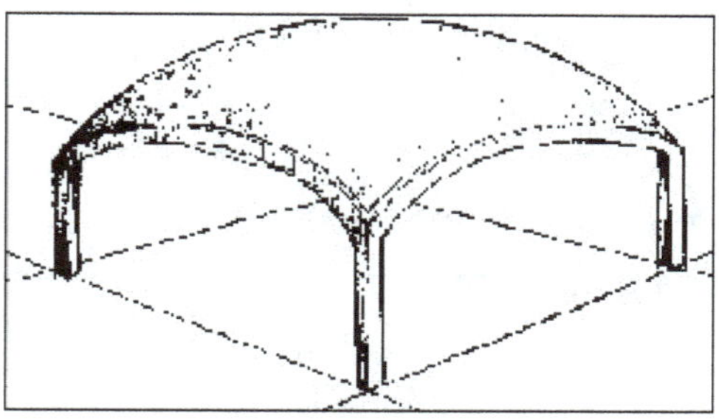

Translation thin shell structure.

Wrapped Surfaces

Warped surfaces have a great advantage for shell structures because they may be formed from straight form boards even though they are surfaces of double curvature.

There are two types which are most useful namely Conoid and hyperbolic paraboloid shell.

Conoid shell structure.

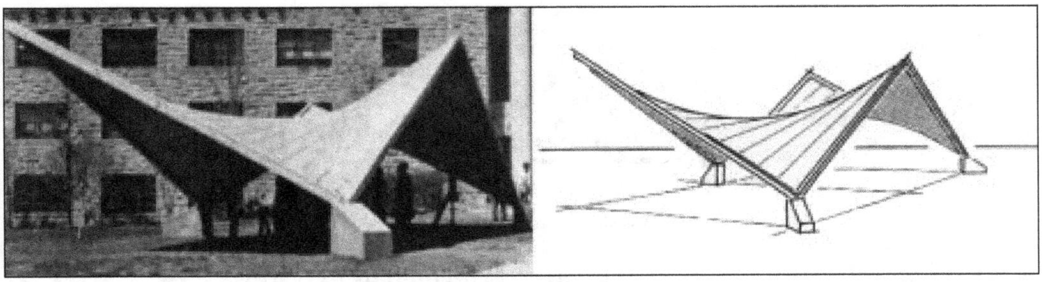

Hyperbolic paraboloid shell.

Combinations

The above shell structures are basic types. So, it is possible to construct different and safer shell structure by combining portion of the basic shell structure.

Intersection shells, barrel shell and folded plate, barrel shell and short shell, barrel shells and domes of revolution, and barrel shells and conoids are all concrete shell combinations. So, numerous combinations can be formed to reach the desired safety and capacity.

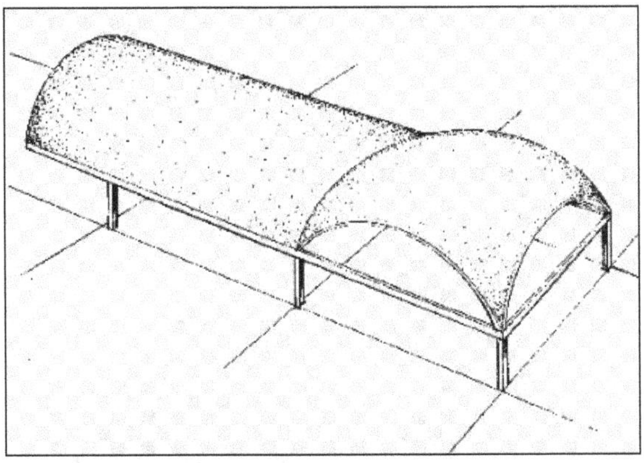

Combination of basic types of shell structures; Dome and Barrel vault.

Arch

An arch is a structure constructed in curved shape with wedge shaped units (either bricks or stones), which are jointed together with mortar, and provided at openings to support the weight of the wall above it along with other superimposed loads. Because of its shape the loads from above gets distributed to supports (pier or abutment).

Different Components of an Arch

The following are the different components of arches and terms used in arch construction:

- Intrados: The inner curve of an arch is called as intrados.

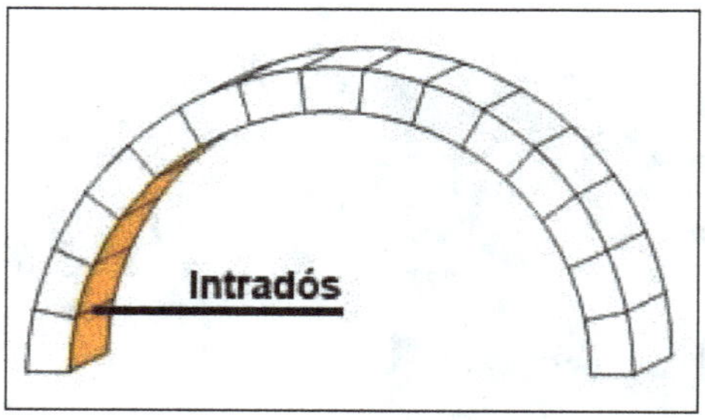

- Extrados: The outer curve of an arch is termed as extrados.

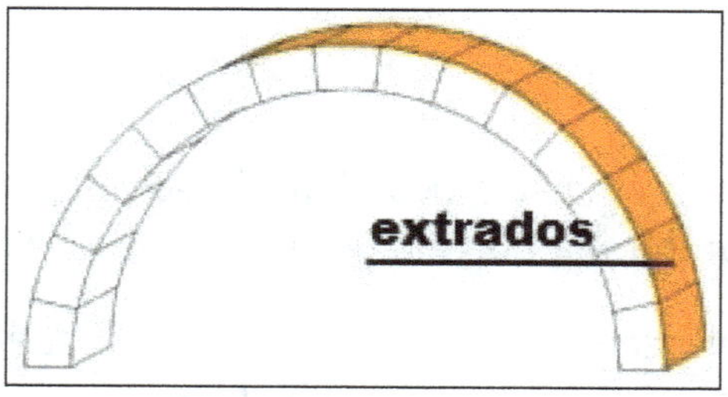

Soffit in an Arch

The inner surface of an arch is called soffit. Soffit and intrados are used synonymously.

Voussoirs

The wedge-shaped units of masonry which are forming an arch are called as voussoirs.

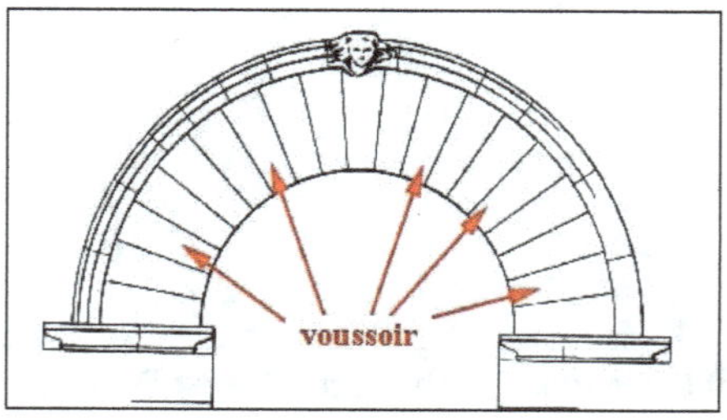

Crown of an Arch

The highest part are peak point of extrados is called crown.

Keystone

The wedge shaped unit which is fixed at the crown of the arch is called keystone.

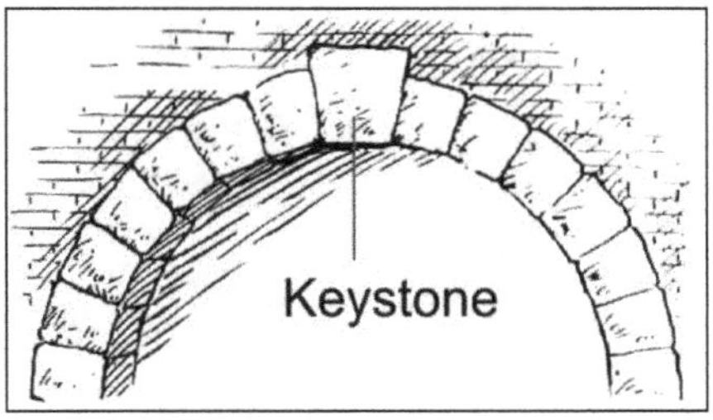

Spandrel in an Arch

If two arches are constructed side by side, then a curved triangular space is formed between the extrados with the base as horizontal line through the crown. This space is called as spandrel.

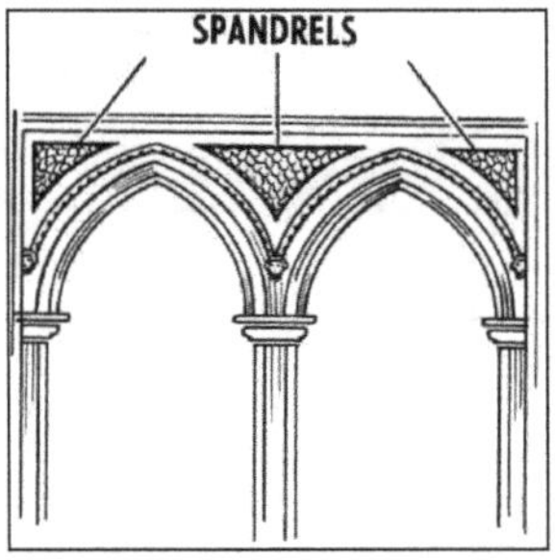

Skew Back

This is an inclined surface or splayed surface on abutment, from which arch curve starts or ends.

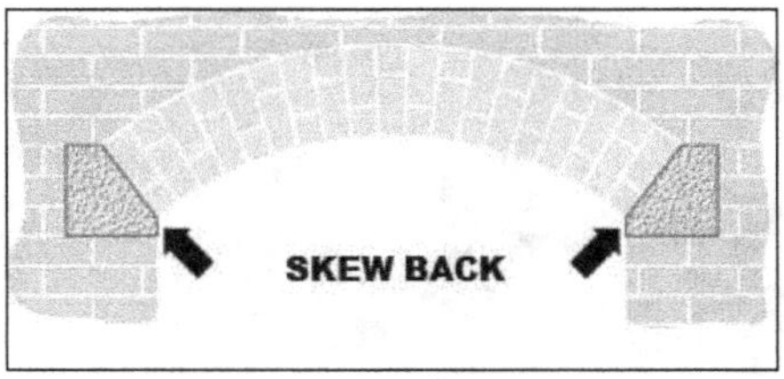

Springing Points

The imaginary points which are responsible for the springing of curve of an arch are called as springing points.

Springing Line

The imaginary line joining the springing points of either ends is called as springing line.

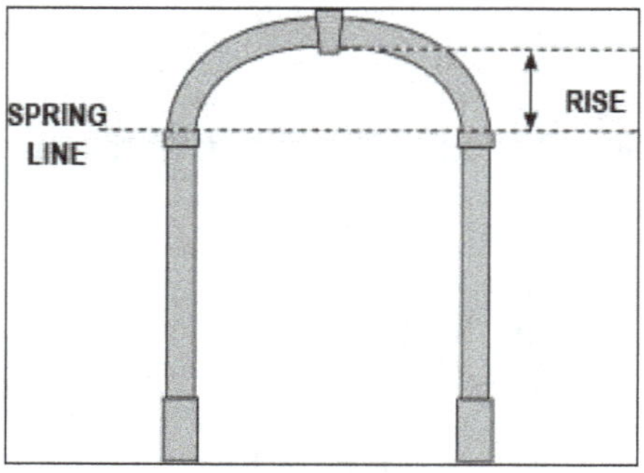

Springer in Arches

The first voussoir at springing level which is immediately adjacent to the skewback is called as springer.

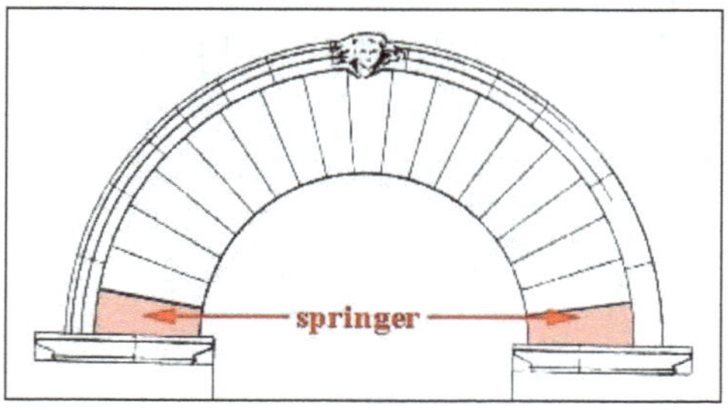

Haunch

The lower half of the arch between the crown and skewback is called haunch. Highlighted area in the figure above is haunch.

Arcade

The row of arches in continuation is called arcade.

Ring

The circular course forming an arch is called as arch ring. An arch may be formed by one ring or combinations of rings.

Impost

The projecting course is provided on the upper part of a pier or abutment to stress the springing line. This course is called impost.

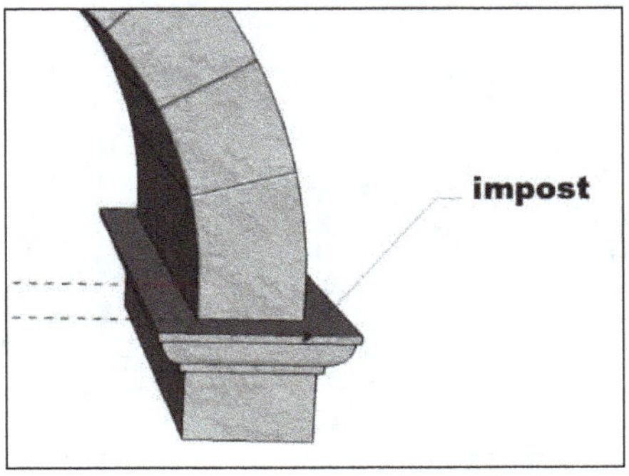

Bed Joints

The joints between the voussoirs are called bed joints. These bed joints are radiate from center of arch. The red color lines in the below figure are bed joints.

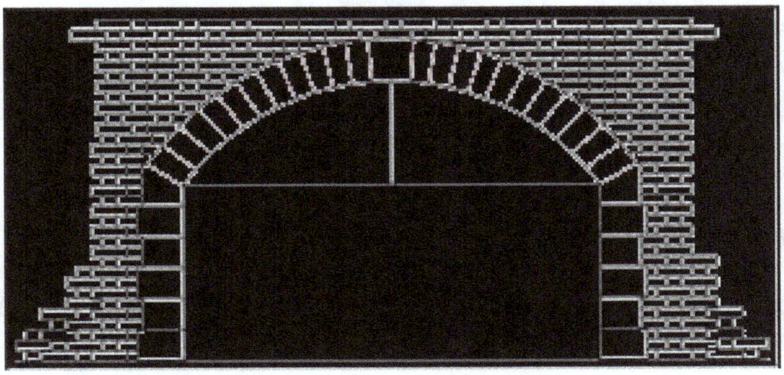

Center of an Arch

The geometrical point based on which the arcs forming intrados of arch, extrados of arch and arch rings are described is the center or striking point.

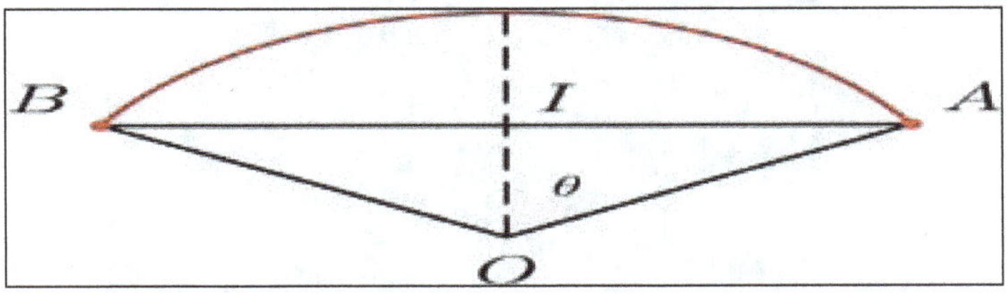

Span of an Arch

The clear horizontal distance between the supports or abutments or piers is termed as span of an arch.

Rise of an Arch

The clear vertical distance between the highest point on the intrados and the springing line is called as rise.

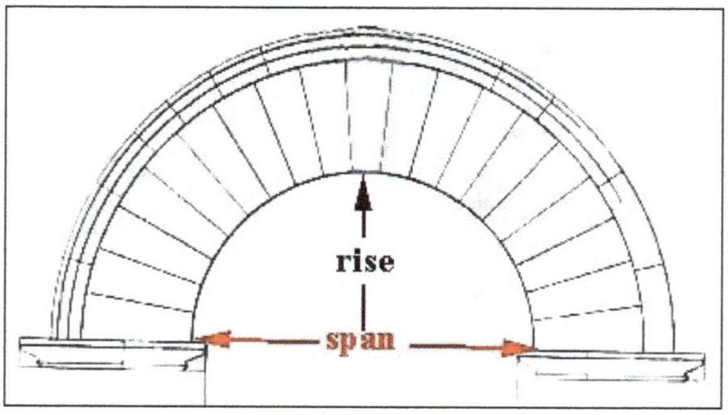

Depth or Height of an Arch

The perpendicular distance between the intrados of arch to the extrados of arch is called depth of an arch or height of an arch.

Thickness of an Arch

This is the breadth of soffit which is measured perpendicular to the front and back faces of an arch. Colored area in the below figure is the thickness of an arch.

Pier and Abutment of an Arch

The intermediate support of an arch is called as pier. The end support of an arch is called as abutment.

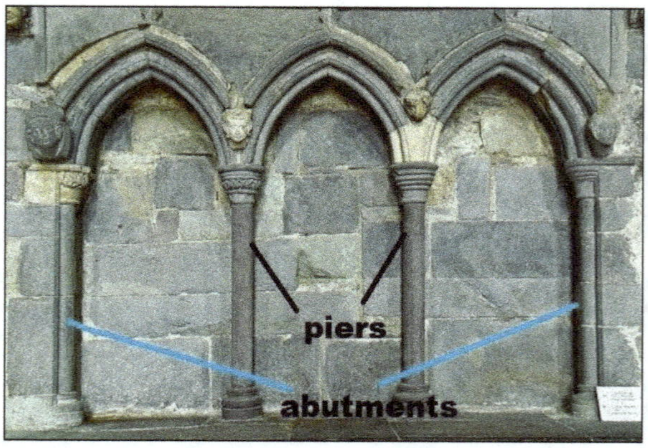

Types of Arches in Construction

Arches are classified based on:

- Shape.
- Number of Centre's.
- Workmanship.
- Materials of construction.

Types of Arches based on shape:

Flat Arch

- For flat arch, the intrados is apparently flat and it acts as a base of equilateral triangle which was formed by the horizontal angle of 600 by skewbacks.
- Even though the intrados is flat but it is given that a slight rise of camber of about 10 to 15 mm per meter width of opening is allowed for small settlements.
- Extrados is also horizontal and flat. These flat arches are generally used for light loads, and for spans up to 1.5m.

Segmental Arch

- This is the basic type of arch used for buildings in which Centre of arch lies below the springing line.
- In segmental arch, the thrust Transferred in inclined direction to the abutment.

Semi-Circular Arch

The shape of arch curve looks like semi-circle and the thrust transferred to the abutments is perfectly vertical direction since skewback is horizontal. In this type of arch, the Centre lies exactly on the springing line.

Horse Shoe Arch

Horse Shoe Arch is in the shape of horse shoe which curves more than semi-circle. This is generally considered for architectural provisions.

Pointed Arch

The other name of pointed arch is Gothic arch. In this type of arch two arcs of circles are met at the apex hence triangle is formed. This may be either isosceles or equilateral.

Venetian Arch

Venetian arch is also pointed arch but its crown is deeper than springing's. It contains four Centre's, all located on the springing line.

Florentine Arch

Intrados of arch is in the shape of semi-circle and rest of the arch is similar to Venetian arch. It has three Centre's, all located on the springing line.

Relieving Arch

Relieving arch is constructed above flat arch or on a wooden lintel to provide greater strength. In case of relieving arch, we can replace the decayed wooden lintel easily without disturbing the stability of structure. The ends of this arch should be carried sufficiently into the abutments.

Stilted Arch

Stilted Arch consists of a semi-circular arch with two vertical portions at the spring-ing's. The Centre of arch lies on the horizontal line through the tops of vertical portions.

Semi-Elliptical Arch

This is a type of arch of semi-ellipse shape and having three or five Centers.

Types of Arches based on Number of Centers

Based on number of centers the arches are classified as:

One-Centered Arches

Segmental, semi-circular, flat, horse-shoe arches and stilted arches are one centered arches. In some cases, perfectly circular arch is provided for circular windows which are called as bull's eye arch is also come under these categories.

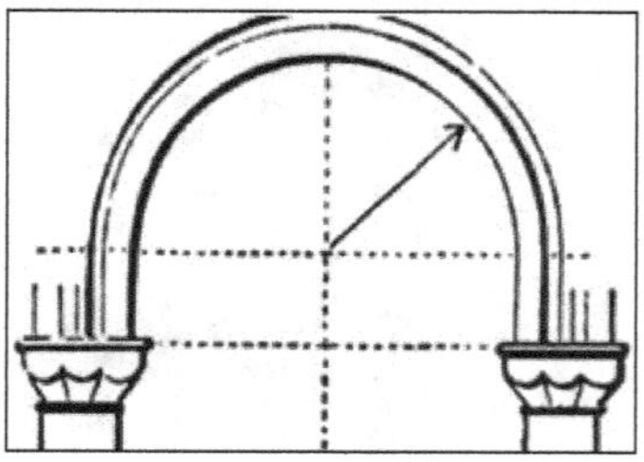

Two Centered Arches

Pointed or gothic or lancet arches are generally come under this type.

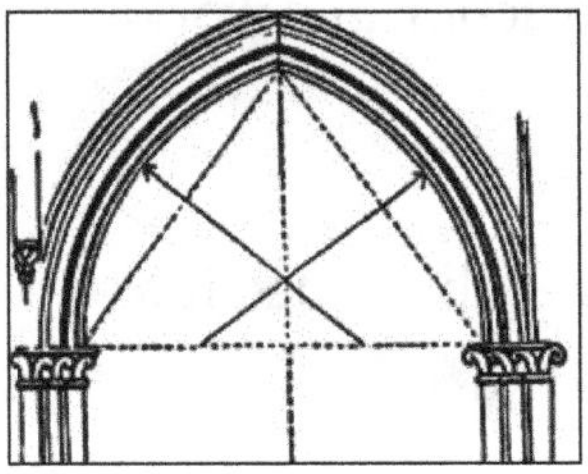

Three Centered Arches

Semi elliptical and Florentine arches are generally having three numbers of centers.

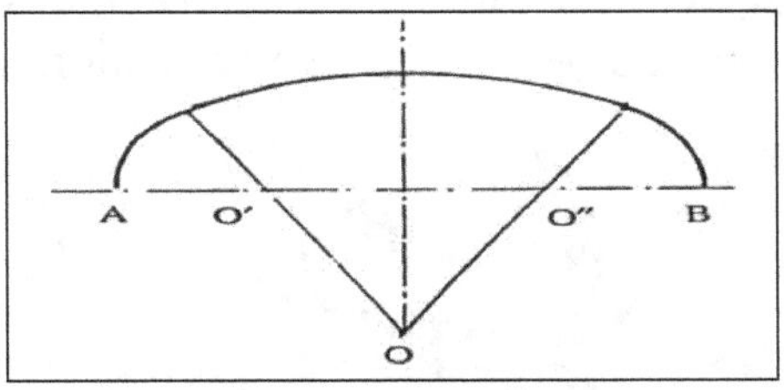

Four Centered Arches

Venetian arch is a typical example for four-centered arch. Tudor arch is also having four centers.

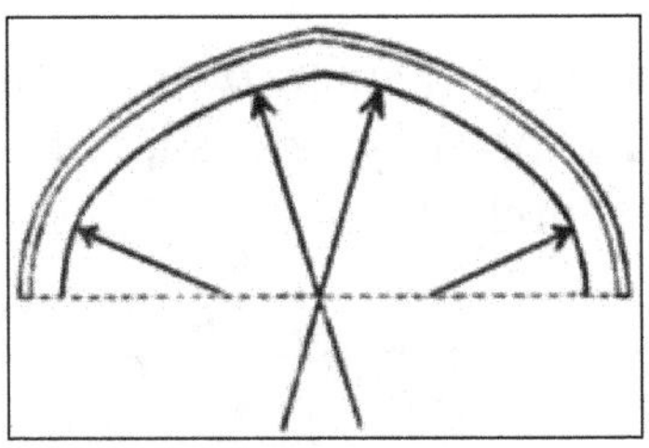

Five centered arches

A good semi-elliptical shape arch contains five centers.

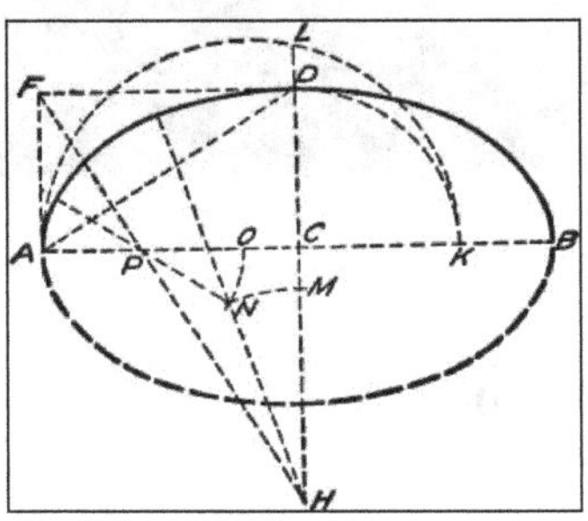

Types of Arches based on Workmanship and Construction Materials

Based on material used for construction and workmanship, arches may be classified as:

Stone Arches

Based on workmanship, these are sub divided into two types. They are:

1. Rubble arches

Rubble arches are very weak and used only for inferior work. These are used up to spans of 1m. These are made of rubble stones which are hammer dressed, roughly to shape and size and fixed in cement mortar. Sometimes these are also used as relieving arches up to a depth of 37.5cm, but these are constructed in one ring. If the depth is more, we can go for two rings in alternate course of headers and stretchers.

2. Ashlar Arches

In this type, the stones are cut to proper shape of voussoirs (a wedge-shaped or tapered stone used to construct an arch) and fully dressed, joined with cement mortar. Ashlar stones are also used to make flat arches.

Brick Arches

Brick arches are also subdivided into:

1. Rough brick arches

These are constructed with ordinary bricks without cutting to the shape voussoirs. The arch curve is provided by forming wedge shaped joints with greater thickness at extrados and smaller thickness at intrados. So, it looks unattractive. That's why it is not recommended for exposed brick works.

2. Axed brick arches

The bricks are cut into wedge shape with the help of brick axe. So, these are roughly dressed in shape and size. Hence, Arch formed by these axed bricks is not very pleasant.

3. Gauged brick arches

In this type arch, bricks are cut to exact shape and size of required voussoir with the help of wire saw. The bricks are finely dressed and these bricks are joined by lime putty. But, for gauged brick arches only soft bricks are used.

4. Purpose made brick arches

The bricks are manufactured, matching with the exact shape and size of voussoirs, to get a very fine workmanship.

Concrete Arches

Concrete arches are of two types:

1. Precast concrete block arches

In Precast concrete block arches the blocks are cast in molds to the exact shape and size of voussoirs. For key stone and skewbacks special molds are prepared. These will give good appearance because of exact shape and size. Cement concrete of 1:2:4 is used.

2. Monolithic concrete block arches

Monolithic concrete block arches are suitable for larger span. These are constructed form cast-in-situ concrete. These may be either plain or reinforced, depending upon the span and magnitude of loading. Form work is used for casting the arch. The curing is done for 2 to 4 weeks.

Catenary

Catenary is a shape formed by the chain when hanging due to downward gravity force acting on the body due to its own self weight, and uniform with respect to the length of the chain. Each foot span of the arch has tension forces acting throughout its length. Shape will have one low point, and is symmetrical about a vertical line at middle of the shape.

Slope of the tangent becomes steeper as the curve moves away from the low point and the angle of the tangent with vertical axis will never get equal to zero.

This shape is said to follow closely the form of the hyperbolic cosine curve. When inverted, the surface of the catenary will be under compression. The inverted catenary shaped arches are constructed in entrance for any buildings. The blocks each support themselves by gravity alone. Each members of the catenary arch are under compression, bearing its own weight.

Mostly the material employed for the arch construction is stones, bricks, and concrete blocks which are strong in compression, and catenary shaped arch is ideal for them. This work does not requires any mortar to be filled in between these bricks for stability,

The catenary shaped domes can be constructed to form a residence building, which can bear a maximum weight with minimal material.

The suspension bridge at the start of the work will be a catenary supporting only its own self weight hanging down being attached to the towers. When the cables are attached to the constructed road deck below it, shape becomes parabolic.

Cable

Cable is also called wire rope which in engineering is either an assemblage of three or more ropes twisted together for extra strength or a rope made by twisting together several strands of metal wire.

The first successful stranded iron wire rope was developed in 1831–34 by Wilhelm Albert, a mining official of Clausthal in the Harz Mountains in Saxony. Even when first tried for hauling and hoisting in his mine, it proved so superior to hemp rope in serviceability and cost that its use soon became widespread in European mining. This stranded wire rope consisted of a number of pieces of wire twisted together to form a wire strand. Several individual wire strands were in turn twisted together to form a wire rope. Albert's first wire rope measured 600 metres (1,968.5 feet) in length and was installed in the Caroline pit of Clausthal in July 1834. Some 18 wire ropes had been constructed by 1838 and put to use in mining operations in the Harz region. Prior to this advancement, wire rope had already been made in the form of a selvage cable—a bundle of individual wires stretched out into a long length and arranged parallel to one another, then bound together and covered with tarred hemp yarns. High-tensile steel wire was introduced during the 1880s, and steel is now the predominant metal used for wire rope.

The manufacture of wire rope is similar to making rope from natural yarns or synthetic filaments. The individual wires are first twisted into strands; six strands (usually), twisted about a core rope, are then laid into the rope. The cores are cord or rope structures made of steel wires; sisal, manila, henequen, jute, or hemp fibres; or polypropylene monofilaments. The function of the core is to provide a firm cushion for positioning the wires in the strands, to maintain a firm rope structure, and to provide some internal lubrication when bending stresses are involved.

Most wire ropes are used in hoisting and hauling operations and in machinery for these purposes, such as cranes, power shovels, elevators, and mine hoists. A flexible rope structure to cope with fast movement and bending stresses is required for most such uses. In other uses, such as support guys and stays, this property is not so much a consideration. Selvage cables composed of parallel wires are ideally suited for the main cables of a suspension bridge. Marine ropes are used for rigging, mooring, and towing.

Cable Structures

Cable structure is a form of long-span structure that is subject to tension and uses suspension cables for support. Highly efficient, cable structures include the suspension bridge, the cable-stayed roof, and the bicycle-wheel roof. The graceful curve of the huge main cables of a suspension bridge is almost a catenary, the shape assumed by any string or cable suspended freely between two points. The cable-stayed roof is supported from above by steel cables radiating downward from masts that rise above roof level. The bicycle-wheel roof involves two layers of tension cables radiating from an inner tension ring and an outer compression ring, which in turn is supported by columns.

References

- Column-and-beam-system-in-construction: basiccivilengineering.com, Retrieved 27 May 2019
- Types-columns-building-construction-24764: theconstructor.org, Retrieved 27 May 2019
- Structural-engg, types-beams-construction-24684: theconstructor.org, Retrieved 27 May 2019
- Concrete-thin-shell-structure-types-forms-6003: theconstructor.org, Retrieved 26 May 2019
- Structures, what-is-an-arch-and-its-components-12081: theconstructor.org, Retrieved 26 May 2019
- Cable-wire-rope, technology: britannica.com, Retrieved 27 March 2019
- Cable-structure, technology: britannica.com, Retrieved 09 August 2019

3

Structural Materials

Structural materials are mainly used for their mechanical properties. These materials include iron, steel, aluminium, timber, bamboo, carbon fibre, adobe, fibre reinforced polymer, etc. This chapter closely examines these different structural materials to provide an extensive understanding of the subject.

Iron

Many ferrous (iron) alloys are used for engineering purposes. Despite this, iron alloys are used in larger quantities than the alloys of any other metal. This arises from the relative cheapness with which steels (an iron alloy) and cast iron can be produced with a variety of useful properties, by variations of composition, heat treatment and heat working. Steel is an alloy of iron and carbon, silicon, phosphorus, manganese and sulphur. Inclusion of nickel and chromium produces stainless steel.

Wrought Iron

Wrought iron is a very low carbon metal made by intensive hand work using small scale production methods. Predominantly, it was manufactured prior to the Patent for the Bessemer process in 1856.

There were numerous methods for producing wrought iron; early iron is often referred to as Charcoal iron, or Bloomery iron, because it was produced in small blooms. In the late 18th century more efficient Puddling furnaces were used to produce Puddled wrought iron. Many other types of production methods were used across the world, all of which produced varying qualities and quantities of iron.

Confusingly, the term 'wrought iron' can also be applied to decorative handmade steel artefacts because wrought means to make by hand (to wrought). However, a more generally accepted description for these handmade steel items is decorative ironwork.

Identifying Wrought Iron

There are four ways in which wrought iron can be identified by using a comparison method:

- Run a file across the iron to see if it cuts easily, do the same to a piece of steel to compare the resistance on the file.

- Using a grinder to observe the spark pattern between the steel and iron; the latter will produce a heavier brighter spark; higher carbon steels will produce a fine duller spark.
- Hold the top of the bar and gently tap it; wrought iron will have a dead sound and steel will ring. High carbon steels will have an even higher ring to them.
- Observe the structure; if it has a wood like grain to it then it is likely to be charcoal iron, however, Puddled wrought iron appears more homogenised like mild steel.

Difference in Performance and Character between Wrought Iron and Steel

Clearly there are visual differences between steel and early wrought iron due to the wood grain effect of the latter. Wrought iron from all periods, and production methods, could take on the same appearance as mild steel if it was repeatedly worked to refine it; furthermore, once painted the visual differences are less apparent to the naked eye.

The structural performance of wrought iron cannot be compared with mild steel, despite the 1938 British Standards because, without reliable records, there is no definitive way of knowing the provenance of the specific wrought iron as found. Analysis from Brunell's iron ship SS Great Britain 1843, by Dr R. Walker in 2002 showed that the mechanical properties were considerably poorer than those tolerated for mild steel.

Corrosion resistance was investigated in 1952 by Chilton and Evans who concluded that even the very best quality wrought iron had not shown 'phenomenally good resistance to corrosion.' earlier investigations by the Hudson Corrosion Committee reported similar findings. There is some evidence that the presence of nickel in the iron ore when smelted may have an influence over corrosion resistance; ores from Sweden and Greece tended to contain higher levels of nickel in the iron ore and showed barriers to corrosion.

Conservation of Wrought Iron

Methods for cleaning and re-coating wrought iron must first be carefully considered. Aggressive forms of cleaning and inappropriate coatings can contribute to further accelerated corrosion. Cardiff University are examining the effects of cleaning and re-coating.

This enquiry should inform decisions regarding the most appropriate methods for the longevity of exposed ironwork.

Corrosion generally takes place at the crevices where the iron is joined. Crevices tend to put stresses on the coating at these junctions causing water to ingress and be trapped, sometimes by the coating itself. Current thinking advises the use of flexible or mobile protective coating to accommodate this expansive action.

In some cases it may be beneficial to control corrosion rather than attempting to stop it. Cardiff University has been contributing to the conservation methods used to preserve Brunel's SS Great Britain using one such method.

Availability of Wrought Iron and Alternatives for Consideration

Wrought iron is still available as recycled iron; the last production of virgin wrought iron from iron ore was in 1974. Good conservation will consider repairs in wrought iron, mild steel, weathering steel (CORTEN), pure iron and other ferrous and non-ferrous metals. The specification should be informed by an experienced professional who can explore all the options and provide good justification for the choice.

Cast Iron

Cast iron is obtained from the pig-iron which is re-melted with coke and limestone. Pig iron is nothing but impure iron which is obtained from the iron ore. Cast iron has lot of engineering properties so, that it can be used in many ways like for sanitary fittings, rail chairs, casting molds etc.

Manufacturing of Cast Iron

Cast iron in manufactured from re-melting process. This process takes place in a furnace called couple furnace. The furnace is 5 meters in height and cylindrical in shape with 1 m diameter.

The raw materials pig iron, coke and lime stone are entered from the charging door of chamber which provided at the top. Air blast is introduced into the chamber using air blast inlet which removes the impurities present in the pig iron.

Hence, pure cast iron is obtained from the bottom out let and it is poured into molds of required shape. These molds are called as cast iron castings.

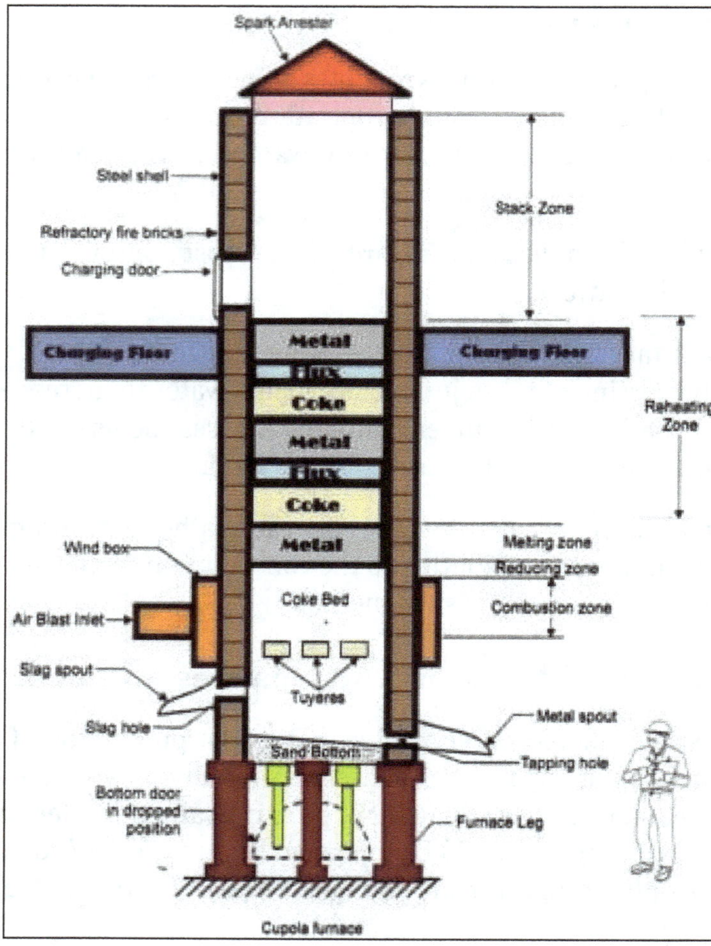

Types of Cast Iron used as a Building Material in Construction Works

Following are the types of cast iron used as a building material in construction and their uses:

Grey Cast Iron

As the name suggests, it is grey in color. It has coarse crystalline structure. Its melting point is very low thus it has weak strength and it is only used for casting purposes.

Malleable Cast Iron

Malleability is the property which helps the materials to undergo any shape without breaking or cracking. Hence, malleable cast iron is used for making many types of materials. It has good corrosive resistance.

Its manufacturing process involves two steps. In the first step, it is casted and cooled as ordinary cast iron and then again it is heated to 1050 °C and soaked in water for long period (several hours or days).

Hence, carbon content is slightly reduced and graphite content is precipitated as temper carbon. This reduces the brittleness of cast iron. So, it can be worked easily using machines. It is used for making pipe fittings, fastenings, automobiles etc.

Malleable Cast Iron Fittings Black Range.

Mottled Cast Iron

Mottled iron is the medium stage cast iron which properties are in between grey cast iron and white cast iron. It has Small amount of graphite in its composition. So, mottled type fractures are developed in its micro structure.

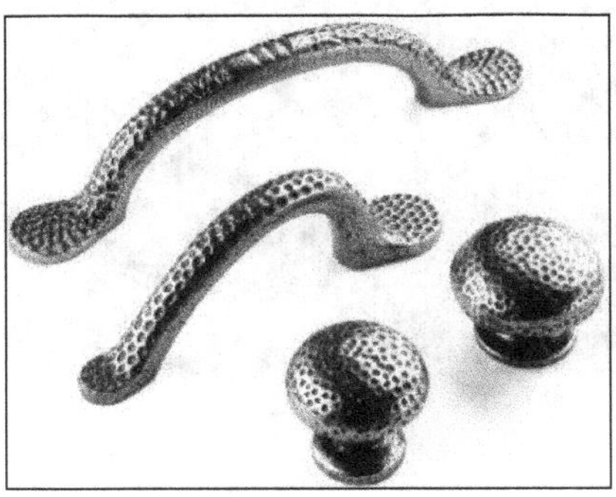

Toughened Cast Iron

Toughened cast iron is the combination of cast iron and wrought iron. To obtain this wrought iron scrap and cast iron melted together. The composition of wrought iron is about 0.15 to 0.25 weight of the cast iron.

White Cast Iron

It is in silver color. Its melting point is high so, strength wise it is better but not used for delicate casting purposes. Because of its heavy strength, it cannot be used easily.

Ductile Cast Iron

Ductile cast iron is also called as spheroidal graphite iron. Its manufacturing process is very easy compared to other types.

Its manufacturing process consists manganese treatment which helps to increase the carbon content and opposes the formation of graphite in flaky form. It has very good engineering properties than malleable cast iron.

Ductile cast iron has very good corrosion resistance, high strength and durability. So, usage of ductile iron dominates the other types. It is used for making sewer pipes, water conveying pipes etc.

Chilled Cast Iron

Chilled cast iron consists two layers of which one layer has white cast iron properties and other one has grey cast iron properties. This type of iron is used for casting process in which grey cast iron layer is provided in inner surface and white cast iron layer is provided as outer surface. Hence the casting molds serve longer. Machine parts are also made using chilled cast iron.

Composition of Different Types of Cast Iron

Different types of cast iron are composed of different quantities of constituents shown in table:

Cast Iron type	Percentage of constituents (%)				
	Carbon	Silica	Manganese	Sulphur	Phosphorus
Grey cast iron	2.5-4.0	1-3	0.2-1	0.02 -0.25	0.02-1
White cast iron	1.8-3.6	0.5-1.9	0.25-0.8	0.06-0.2	0.06-0.2
Ductile cast iron	3-4	1.8-2.8	0.1-1.0	0.01-0.03	0.01-0.1
Malleable cast iron	2.0-2.9	0.9-1.9	0.15-1.2	0.02-0.2	0.02-0.2
Mottled cast iron	2.5-4.0	1.0-3.0	0.2-1.0	0.01-0.03	0.01-0.1

Properties of Cast Iron as a Building Material

The properties of cast iron are as follows:

- Good corrosion resistance, so it can be used for water carrying pipes etc.
- Does not get attracted to magnet.
- Specific gravity is 7.5.
- Melting point is about 1250 °C.
- Tensile strength is about 150 N/mm² and compressive strength is about 600 N/mm².so, it is good in compression.
- It becomes soft when placed in salt water and it shrinks on cooling.
- It cannot be useful for forging work because of lack of plasticity.

Uses of Cast Iron in Building Construction

Cast iron can be used for making different tools, materials etc. as described:

- Many types of sanitary fittings like manholes, sewer pipes, water pipes, cisterns are manufactured using cast iron.

- Metal columns and column bases can be made using cast iron.
- Casting molds for making metal staircases, lamp posts, gates etc. are made using cast iron.
- Rail chairs and carriage wheels are manufactured.
- Several types of agricultural implements can be made.
- Machinery parts can be manufactured but shock cannot be resisted by cast iron.

Steel

Steel is an alloy of iron and carbon in which the carbon content ranges up to 2 percent (with a higher carbon content, the material is defined as cast iron). By far the most widely used material for building the world's infrastructure and industries; it is used to fabricate everything from sewing needles to oil tankers. In addition, the tools required to build and manufacture such articles are also made of steel. As an indication of the relative importance of this material, in 2013 the world's raw steel production was about 1.6 billion tons, while production of the next most important engineering metal, aluminum, was about 47 million tons. The main reasons for the popularity of steel are the relatively low cost of making, forming, and processing it, the abundance of its two raw materials (iron ore and scrap), and its unparalleled range of mechanical properties.

Use of Steel in Buildings and Infrastructure

The possibilities for using steel in buildings and infrastructure are limitless. The most common applications are listed below:

For Buildings

- Structural sections: these provide a strong, stiff frame for the building and make up 25% of the steel use in buildings.
- Reinforcing bars: these add tensile strength and stiffness to concrete and make up 44% of steel use in buildings. Steel is used because it binds well to concrete, has a similar thermal expansion coefficient and is strong and relatively cost-effective. Reinforced concrete is also used to provide deep foundations and basements and is currently the world's primary building material.
- Sheet products: 31% is in sheet products such as roofing, purlins, internal walls, ceilings, cladding, and insulating panels for exterior walls.

- Non-structural steel: steel is also found in many non-structural applications in buildings, such as heating and cooling equipment and interior ducting.
- Internal fixtures and fittings such as rails, shelving and stairs are also made of steel.

For Infrastructure

- Transport networks: steel is required for bridges, tunnels, rail track and in constructing buildings such as fueling stations, train stations, ports and airports. About 60% of steel use in this application is as rebar and the rest is sections, plates and rail track.
- Utilities (fuel, water, power): over 50% of the steel used for this application is in underground pipelines to distribute water to and from housing, and to distribute gas. The rest is mainly rebar for power stations and pumping houses.

Stainless Steel

Stainless steel refers to a group of corrosion-resistant steel alloys containing about 10.5% chromium. This chromium in the stainless steel reacts with the oxygen in the atmosphere and produces a thin and inert layer of chromium-rich oxide on its surface. Due to this layer, stainless steel is corrosion resistant. Unlike other protective coatings such as paint or galvanising, this protective film of chromium-rich oxide continues to protect the steel even if abrasion damages it. This film tends to re-form even after being damaged.

Stainless steel is widely used for providing support and strength to the structure and is extremely durable. However, from an architectural point of view, stainless steel is also ideal for creating style. For any concrete structure, enhancements can be easily made using stainless steel. Moreover, when constructing a building, it needs a solid structure to support the entire building. Many important parts of a building including wind posts, lintels, and masonry are made from stainless steel. Stainless steel strips, steel wires, and steel plates are widely used to manufacture these fixings. In construction, stainless steel is also used for:

- Flexible Design: Structural steel encourages design creativity. No matter how complex the design, it can easily accomplish with the help of structural steel.
- It improves the aesthetic appeal: Steel has a natural beauty, and the architects can easily expose this beauty in the design of their structures for emphasising the grace, strength, and transparency of the frame. No other material allows the application of degrees of expression and creativity as structural steel.
- It is cost effective: The substitutes of stainless steel are extremely costly. So, without stainless steel, there would be a rise in prices. Moreover, without

stainless steel, it would be a bit tough to produce aesthetically pleasing structures and buildings that are resistant to corrosion.

Reasons to use Stainless Steel

Stainless steel is widely used for construction purpose all over the globe. This is because stainless steel offers a number of benefits such as:

- It is resistant to corrosion.
- It is highly ductile and offers superior strength.
- It has excellent high and low-temperature properties.
- It is resistant to staining.
- It has an aesthetic surface finish.
- It is 100% recyclable.

Stainless Steel is widely used within the construction industry.

Chosen for use within the construction industry for various reasons including the fact that it has a high resistance to corrosion as well as high and low temperatures the majority of stainless steel is fairly easily fabricated. It comes in a whole range of surface finishes which are easy to maintain.

Stainless Steel is used to provide support and strength to a structure but in regards to architecture steel is used for construction work which not only needs to fulfil its purpose and be durable but stainless steel is ideal for also creating style too. Enhancements to buildings can be made from stainless steel which if they were produced from another material they would prove extremely costly.

When a building is being constructed it needs an adequate structure to it in order for it to have support. Many parts of the structure are made from stainless steel including lintels, wall starters, windposts and masonary supports. These are manufactured form stainless steel strip, stainless steel plate and stainless steel wire. In addition a range of fixings can be used including wall ties, tying wire, screws and nails all manufactured form stainless steel.

Some of the most popular uses of stainless steel in construction are:

- Architectural Cladding.
- Handrails.
- Drainage and Water Systems.
- Wall Support Products.

- Roofing.
- Structure.
- Fixings.

Without stainless steel in construction we would probably see a rise in prices to build because the other substitutes are more expensive.

Concrete

In construction, concrete is a structural material consisting of a hard, chemically inert particulate substance, known as aggregate (usually sand and gravel), that is bonded together by cement and water.

Concrete: Construction workers pouring concrete.

Among the ancient Assyrians and Babylonians, the bonding substance most often used was clay. The Egyptians developed a substance more closely resembling modern concrete by using lime and gypsum as binders. Lime (calcium oxide), derived from limestone, chalk, or (where available) oyster shells, continued to be the primary pozzolanic, or cement-forming, agent until the early 1800s. In 1824 an English inventor, Joseph Aspdin, burned and ground together a mixture of limestone and clay. This mixture, called portland cement, has remained the dominant cementing agent used in concrete production.

Aggregates are generally designated as either fine (ranging in size from 0.025 to 6.5 mm [0.001 to 0.25 inch]) or coarse (from 6.5 to 38 mm [0.25 to 1.5 inch] or larger). All aggregate materials must be clean and free from admixture with soft particles or vegetable matter, because even small quantities of organic soil compounds result in chemical reactions that seriously affect the strength of the concrete.

Concrete is characterized by the type of aggregate or cement used, by the specific qualities it manifests, or by the methods used to produce it. In ordinary structural concrete, the character of the concrete is largely determined by a water-to-cement ratio. The lower the water content, all else being equal, the stronger the concrete. The mixture must have just enough water to ensure that each aggregate particle is completely surrounded by the cement paste, that the spaces between the aggregate are filled, and that the concrete is liquid enough to be poured and spread effectively. Another durability factor is the amount of cement in relation to the aggregate (expressed as a three-part ratio—cement to fine aggregate to coarse aggregate). Where especially strong concrete is needed, there will be relatively less aggregate.

The strength of concrete is measured in pounds per square inch or kilograms per square centimetre of force needed to crush a sample of a given age or hardness. Concrete's strength is affected by environmental factors, especially temperature and moisture. If it is allowed to dry prematurely, it can experience unequal tensile stresses that in an imperfectly hardened state cannot be resisted. In the process known as curing, the concrete is kept damp for some time after pouring to slow the shrinkage that occurs as it hardens. Low temperatures also adversely affect its strength. To compensate for this, an additive such as calcium chloride is mixed in with the cement. This accelerates the setting process, which in turn generates heat sufficient to counteract moderately low temperatures. Large concrete forms that cannot be adequately covered are not poured in freezing temperatures.

Concrete: Workers using trowels to smooth freshly poured concrete.

Concrete that has been hardened onto imbedded metal (usually steel) is called reinforced concrete, or Ferro concrete. Its invention is usually attributed to Joseph Monier, a Parisian gardener who made garden pots and tubs of concrete reinforced with iron mesh; he received a patent in 1867. The reinforcing steel, which may take the form of rods, bars, or mesh, contributes tensile strength. Plain concrete does not easily withstand stresses such as wind action, earthquakes, and vibrations and other bending forces and is therefore unsuitable in many structural applications. In reinforced concrete, the tensile strength of steel and the compressional strength of concrete render a member capable of sustaining heavy stresses of all kinds over considerable spans. The fluidity of the concrete mix makes it possible to position the steel at or near the point where the greatest stress is anticipated.

Another innovation in masonry construction is the use of prestressed concrete. It is achieved by either pretensioning or posttensioning processes. In pretensioning, lengths of steel wire, cables, or ropes are laid in the empty mold and then stretched and anchored. After the concrete has been poured and allowed to set, the anchors are released and, as the steel seeks to return to its original length, it compresses the concrete. In the posttensioning process, the steel is run through ducts formed in the concrete. When the concrete has hardened, the steel is anchored to the exterior of the member by some sort of gripping device. By applying a measured amount of stretching force to the steel, the amount of compression transmitted to the concrete can be carefully regulated. Prestressed concrete neutralizes the stretching forces that would rupture ordinary concrete by compressing an area to the point at which no tension is experienced until the strength of the compressed section is overcome. Because it achieves strength without using heavy steel reinforcements, it has been used to great effect to build lighter, shallower, and more elegant structures such as bridges and vast roofs.

In addition to its potential for immense strength and its initial ability to adapt to virtually any form, concrete is fire resistant and has become one of the most common building materials in the world.

Reinforced Concrete

Reinforced concrete is a type of concrete in which steel is embedded in such a manner that the two materials act together in resisting forces. The reinforcing steel—rods, bars, or meshes—absorbs the tensile, shear, and sometimes the compressive stresses in a concrete structure. Plain concrete does not easily withstand tensile and shear stresses caused by wind, earthquakes, vibrations, and other forces and are therefore unsuitable in most structural applications. In reinforced concrete, the tensile strength of steel and the compressive strength of concrete work together to allow the member to sustain these stresses over considerable spans. The invention of reinforced concrete in the 19th century revolutionized the construction industry, and concrete became one of the world's most common building materials.

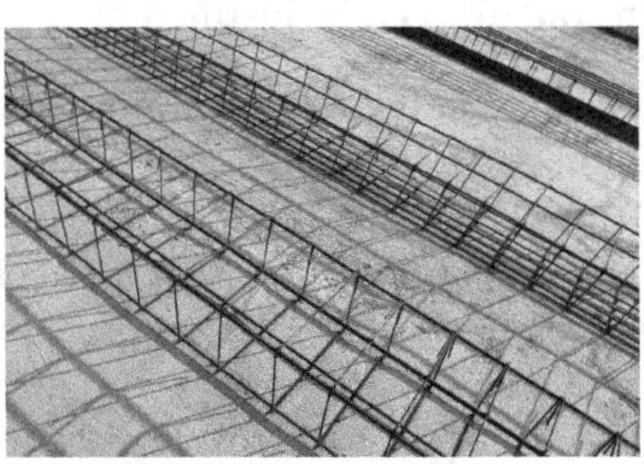

Reinforced concrete: Wire reinforcements placed before the pouring of concrete.

Prestressed Concrete

Prestressed concrete is a system into which internal stresses are deliberately induced without any form of external loads to improve its performance. The internal stresses induced in the concrete structure are used to counteract the stresses coming from the external load application.

Concept of Prestressing

Prestressing is the application of an initial load on the concrete structure, so that the structure is able to counteract or withstand the stresses coming due to the service loads. The concept can be clearly understood by the example of a barrel.

A barrel used in olden periods to transports liquids and grains are tightly bound by metal bands as shown in figure below. These metal bands are fitted so tightly that it creates a hoop compression around the barrel. When this barrel is filled with liquid, it exerts hoop tension.

The hoop compression that is created by the metal bands helps to counteract the hoop tension created by the fluid within. This is a prestressing system.

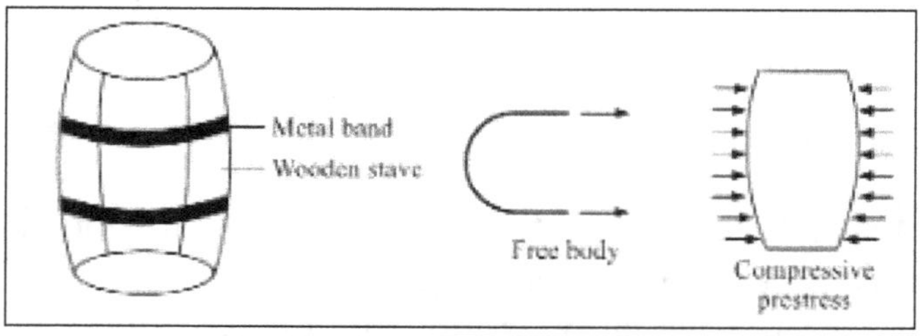

Prestressing Concept in a Barrel.

Similarly, effective internal stresses are induced into the concrete by means of tensioned steel bars before the concrete structure is subjected to any service loads. This stress counteracts with external stresses.

Need for Prestressing Concrete

The need for prestressing in concrete can be justified by the following issue:

- Concrete is weak in tension and strong in compression. This is a weak point of concrete that results in early flexural cracks mainly in flexural members like beams and slabs. To prevent this, the concrete is induced with compressive stress deliberately (prestressing) and this stress counteracts with the tensile stress the structure is subjected to during service condition. Hence the chances of flexural cracks are reduced.
- The pre-compression that is induced as a part of prestressing helps to enhance the bending capacity, the shear capacity and the torsional capacity of the flexural

members.

- A compressive prestressing force can be applied concentrically or eccentrically in the longitudinal direction of the member. This prevents cracks at critical midspan and supports at service load.
- A prestressed concrete section behaves elastically.
- The full capacity of the concrete in compression can be used over entire depth under full loading in the case of prestressed concrete.

Methods of Prestressing

The prestressing can be performed by two methods:

Pretensioning

In the pretensioning method, the stress is induced by initially tensioning the steel tendons. These are wires or strands that are tensioned between the end anchorages. After this tensioning process, the concrete casting is performed.

Once the casted concrete has hardened sufficiently, the end anchorages arranged are released. This releasing transfers the prestress force to the concrete. The bond between the concrete and the steel tendons facilitates this stress transfer.

As shown in figure, the tendons that are protruding at the ends are cut and a finished look is achieved. In order to induce prestress force in the pre-tensioning method, a large number of tendons and wires are used. This arrangement hence demands a large area of surface contact to make the bond and stress transfer possible.

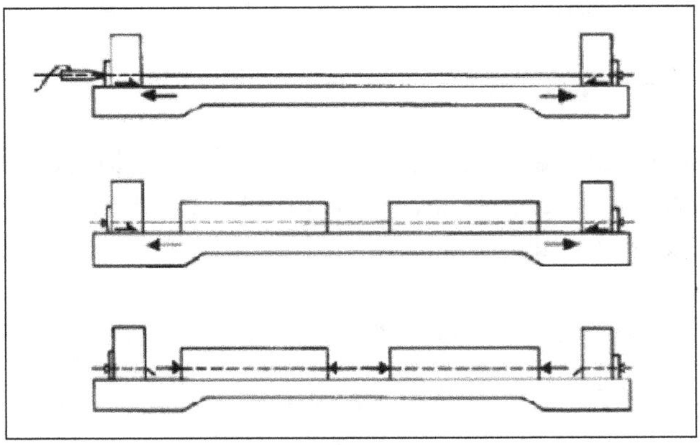

The process of Pre-tensioning.

Post Tensioning

The procedure in post-tensioning is depicted in the figure. Here, the steel is prestressed only after the beam is cast, cured and attain strength to take the prestress. Within the sheathing, the concrete is cast. For the passage of steel cables, ducts are formed in the concrete.

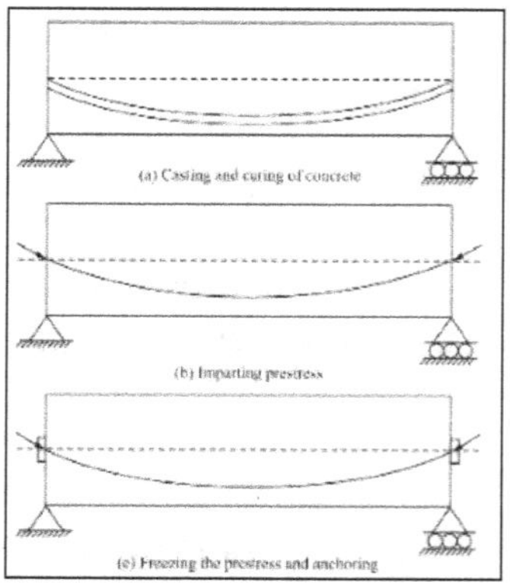

The process of Post-tensioning.

Once the casted concrete hardens completely, the tendons are tensioned. One end of the tendon is anchored and the other end is tensioned. In some cases, the tensioning can be performed from either side and anchored subsequently.

Once the prestressing is complete, there is space between the tendons and the duct. This leads to:

Bonded Construction

In bonded construction, the space between the duct and the tendon is filled with cement grout. The grouting process helps the steel to resist corrosion to a large extent. The ultimate strength is increased as this method increases the resistance to live loads acting. The grout mixture is cement and water combined with or without admixture. No sand is used in this grout.

Unbonded Construction

If no grout is used to fill the space between the duct and tendon, it is called as unbonded construction. Here, the steel is galvanized to protect from corrosion. A waterproofing material is used for galvanizing.

Advantages of Prestressed Concrete

The major advantages of Prestressed Concrete are:

- The prestressing of concrete by using high tensile steel improve the efficiency of the materials.
- The prestressing system works for a span greater than 35m.
- Prestressing enhance shear strength and fatigue resistance of concrete.

- Dense concrete is provided by prestressing systems thus improving the durability.
- Best choice for the construction of sleek and slender structures.
- Prestressing helps to reduce the dead load of the concrete structure.
- Prestressed concrete remains uncracked even an at service load condition which proves the structural efficiency.
- Composite construction by using the prestressed concrete unit and cast-in-unit derives the economical structure.

Disadvantages of Prestressed Concrete

- Higher material costs.
- Prestressing is an added cost.
- Formwork is more complex than for RC (flanged sections, thin webs) – thus, precast not as ductile as RC.

Aluminium

Aluminium is widely used in building because of its intrinsic properties of lightness and corrosion resistance.

Aluminum is used in external facades, roofs and walls, in windows and doors, in staircases, railings, shelves, and other several applications.

Aluminium in Building Construction

There are many benefits that aluminum offers to the construction industry:

1. Alloys: Pure aluminum is a low-strength metal and consequently not suitable for building applications but thanks to the addition of alloying elements such as copper, manganese,

magnesium, zinc etc. and thanks to specific production processes, it changes its physical and mechanical properties to meet requirements of a large number of applications.

2. Durability: Aluminum alloys for Building are resistant to water, corrosion and immune to the harmful effects of UV rays, thus ensuring a lasting endurance.

3. Low maintenance costs: Aluminum does not require any special kind of maintenance, whether it is raw or lacquered aluminum.

4. Finishes: Aluminum can be anodized or lacquered in any color, so it's possible to get the most varied effects and thus meet the designer's decorative needs. Aluminum treatments can increase the durability of the material and its corrosion resistance.

5. Reflective properties: Aluminum is widely used for light management: its reflective properties help to reduce energy consumption for lighting and heating.

For example, it's possible to reduce the use of air conditioning in summer season by using aluminum shielding devices.

6. Aluminum is not combustible: Aluminum does not burn and is therefore it's classified as non-combustible material (A1 fire reaction class).

Nevertheless, aluminum alloys melt at about 650 ° C, but without releasing any harmful gases. And so, more and more often, the outer covers and external surfaces of industrial structures (and not) are made with thin aluminum panel finishes which are destined to merge only in case of fierce fire, thus allowing heat and smoke to escape and reducing damage Caused by the fire.

7. Aluminum is eco-friendly.

Composites

Composite material is also called composite which is a solid material that results when two or more different substances, each with its own characteristics, are combined to create a new substance whose properties are superior to those of the original components in a specific application. The term composite more specifically refers to a structural material (such as plastic) within which a fibrous material (such as silicon carbide) is embedded.

The remarkable properties of composites are achieved by embedding fibres of one substance in a host matrix of another. While the structural value of a bundle of fibres is low, the strength of individual fibres can be harnessed if they are embedded in a matrix that acts as an adhesive, binding the fibres together and lending solidity to the material. The rigid fibres impart structural strength to the composite, while the matrix protects the fibres from environmental stress and physical damage and imparts thermal stability to them. The fibre-matrix combination also reduces the potential for a complete fracture; if one fibre fails the crack may not extend to other fibres, whereas a crack that starts in a monolithic (or single) material generally continues to propagate until that material fails.

Most conventional composites resemble plywood in that they are built in thin layers, each of which is reinforced by long fibres laid down in a single direction. Such materials exhibit enhanced strength only along the direction of the fibres. To produce composites that are strong in all directions, the fibres are woven into a three-dimensional structure in which they lie along three mutually perpendicular axes.

The structural component of a composite may consist of fibres made of glass or carbon-graphite, shorter “whiskers” made of silicon carbide or aluminum oxide, or longer tungsten-boron filaments. The matrix material may be an epoxy resin or other high-temperature plastic, aluminum or some other metal, or a ceramic such as silicon nitride. Fibreglass-reinforced plastic is the best-known composite and has found wide application in both household goods and industrial products. Composites are of greatest use in the aerospace industry, however, where their stiffness, lightness, and heat resistance make them the materials of choice in reinforcing the engine cowls, wings, doors, and flaps of aircraft. Composite materials are also used in rackets and other sports equipment, in cutting tools, and in certain parts of automotive engines.

Timber

Wood is one of the most used natural building materials in the world. A number of valuable properties such as low heat conductivity, small bulk density, relatively high strength, amenability to mechanical working etc. makes wood as famous building material.

Timber can be used in most economical way without wasting any of the derivatives of it. Even the saw dust obtained during wood sawing can also be used to make fiber boards, paper etc.

Types of Trees for Timber Production

The trees are classified into following types based on their mode of growth:

Exogenous

Exogenous trees are outward growing trees. Horizontal section of such tree contains several rings which are nothing but annual rings. These rings can be used to predict the age of tree. Most of the exogenous trees are useful for many engineering purposes.

Exogenous trees are sub classified into following types:

- Conifers,
- Deciduous.

Exogenous Tree.

Conifers

Conifers are nothing but soft wood producing trees which are also called as ever green trees. The timber of these trees is light colored, light in weight, low dense and poor against fire.

Examples: Pine, Fir, redwood, spruce, deodar, cedar etc.

Deciduous

Deciduous trees are hard wood producing trees. The leaves of this type of trees are generally broad in size and they fall in autumn and grow in spring. Deciduous trees are most suitable for constructional purposes. Timber of deciduous trees is dark colored, dense, heaviest and good against fire. Examples: Maple, Mahogany, Oak, teak, walnut, babul etc.

Endogenous

Endogenous trees are inward growing trees which contains fibrous mass in their longitudinal section. The timber from these trees is useful in some limited engineering purposes. Examples: bamboo, palm, cane etc.

Endogenous tree.

Structure of a Tree

The structure of tree can be divided in to two categories as follows:

Macrostructure

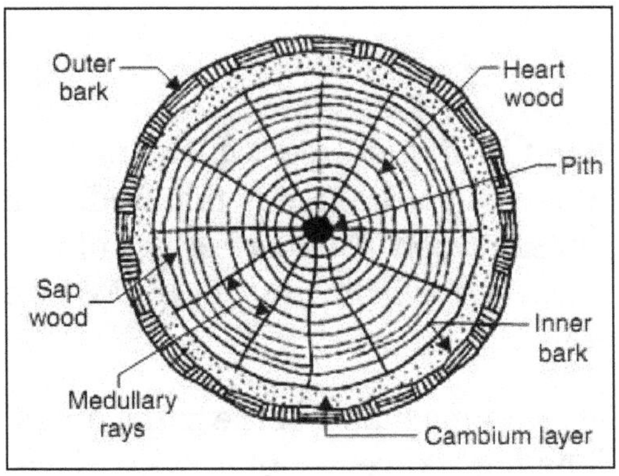

Macrostructure of Tree.

The structure of a tree which is visible to the naked eye is called macro structure of tree. Macrostructure of tree contains following components:

- Pith: The core part or innermost part of the tree is called as pith. It contains cellulose tissues which are helpful for the growth of plant during its young age.
- Heart wood: Heart wood is the portion around the pith which is dark in color and contains some annual rings in it. It is very hard and provides rigidity to the tree. Heart wood is used for several engineering purposes because of its strength and durability.
- Sap wood: Sap wood contains outer annual rings. This indicates the recent growth of tree and is light in color. It contains sap which helps in the growth of tress.
- Cambium Layer: Cambium layer contains sap which will turn into sap wood after some time. It should not expose to atmosphere otherwise the tree may dead.
- Inner bark: The protecting layer of cambium layer is known as inner bark.
- Outer bark: The outermost layer of the tree section is called outer bark or cortex. It contains cells of wood fiber.
- Medullary rays: The rays extending from pith to cambium layer are known as medullary rays. These rays hold the annual rings of sap wood and heart wood together.

Microstructure

Micro structure of a tree can only be visible with great magnifications. It contains cells of different shapes and sizes. These cells are responsible for many actions like nutrients transport to branches from stem, strength of tree etc.

Processing of Timber

Processing of Timber contains following steps:

1. Felling of Trees

Felling of trees is nothing but cutting of trees which are suitable for engineering purposes. Felling should be done when the tree is matured. Then only it contains more amount of heart wood than sap wood. The perfect age of trees for felling varies from 50 to 100 years. The best season for felling of trees is Mid-winter for plain areas and mid-summer for hilly areas.

Firstly a cut is made at the lower most part of the trunk on a side where tree is expected to fell. The cut should be beyond center of gravity of tree. Then provide parallel cut which is exactly opposite to the first cut. Then tie up the tree top with 4 ropes on 4 sides.

Now pull the rope of first cut side and loosen the rope on the opposite side. Using other two ropes swing the tree slowly. Then the tree starts breaking along the cuts and gently falls on the ground. The branches are chopped off; bark is removed and is cut into required sizes.

Felling of a Tree.

2. Seasoning of Timber

Seasoning of timber means removal of moisture content from timber. A newly fell tree contains water up to 50% of its dry weight. The timber contains free moisture and bound moisture.

Free moisture is present in timber as water vapor while bound moisture is present in cell walls. When it is allowed to seasoning, free moisture evaporated first and this point is called as fiber saturation point.

After Fiber saturation point, the timber will shrink on drying which is nothing but evaporation of bound moisture.

There are two methods of seasoning are there namely:

- Natural seasoning.
- Artificial Seasoning.

Seasoning of Timber.

3. Conversion of Timber

Conversion of timber is the process of cutting of timber into required sections. This can be done by using power machines. Skilled persons should be required for economic conversion of timber. The conversion can be done by four types as follows:

- Ordinary sawing: It is the most used and easy method of sawing. The cutting is done through the section of timber piece at perpendicular to the annual rings. Wastage of timber is minimum in this case.

  The obtained planks are not of equal strength. The outer plank contains sap wood and shrinks more while the inner portion planks contain heart wood which shrinks less.

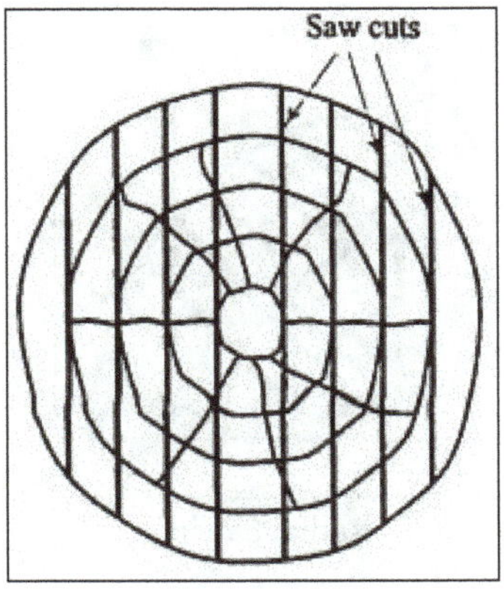

Ordinary Sawing.

- Tangential sawing: In this type of sawing the cuts are tangential to annual rings and they meet each other at right angles. This method is suitable when the annual rings are very distinct with each other.

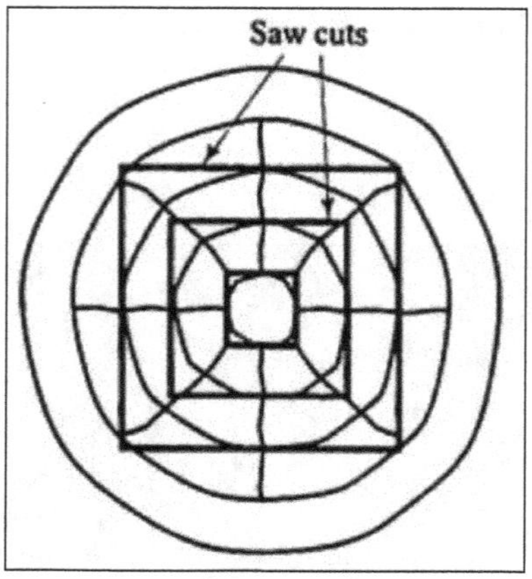

Tangential Sawing.

- Quarter sawing: In quarter sawing, the cuts are made right angles to each other. This is suitable when the timber have no distinct medullary rays.

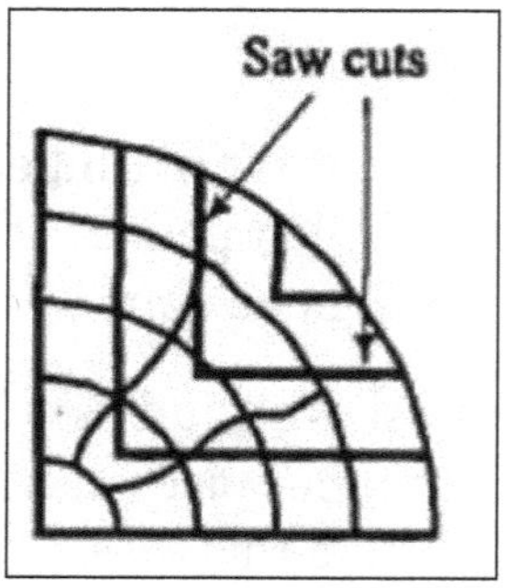

Quarter Sawing.

- Radial sawing: In this type of sawing, the cuts are made parallel to the medullary rays radially. In this method wastage of timber is maximum.

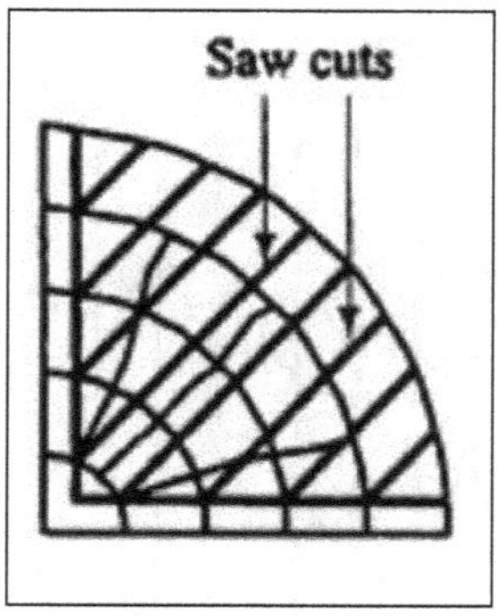

Radial Sawing.

4. Preservation of Timber

The final stage of timber processing is preservation which is carried out to increase the durability of timber and also to resist the attacks of fungi, insects etc. on timber. In general ASCU, Coal tar, Oil paints, Solignum paints etc. are used as preservatives.

Preservation of Timber.

Other Structural Materials

Bamboo

Bamboo as a building material has high compressive strength and low weight has been one of the most used building materials as support for concrete, especially in those locations where it is found in abundance.

Bamboo as a building material is used for the construction of scaffolding, bridges and structures, houses.

Due to a distinctive rhizome-dependent system, bamboos are one of the fastest-growing plants in the world and their growth is three times faster than most other species of plants. They are renewable and extremely versatile resource with multi-purpose usage. Among many uses of bamboo, Housing is one of the major areas applications especially in the wake of residential shortages around the globe.

Bamboo as a building material is conventionally associated with the region of Southeast Asia and South America where climate is best suitable for its cultivation. In many of the nations, bamboo is used to hold up suspension bridges or simply make places of dwelling.

Various Structural Shapes of Bamboo as a Building Material

Bamboos are treated in such a way that they assume desired shapes and structures while they grow.

- Squared cross-section can be obtained by compressing the growing stalk of bamboo within a square section.
- Arch shapes of bamboo can also be created by compressing the bamboo's growth into the desired shape. This would cost lesser than it would to get the same form with normal timber.
- Curved and Flat shapes of bamboo are achieved through traditional techniques like applying heat and pressure.

Methods of Working on Bamboo

For a bamboo to be used as a building material, it must be worked on to create desired shape, bend and length to be used for structural or other purposes.

Following are the different works involved with use of bamboo:

Splitting of Bamboo

The bamboo canes are split into halves or quarter sections using a knife ideal for the job and setting them apart by a wedge. About four or eight segments can be acquired which are used as canes, strips or battens. Canes can be peeled to make strings and ropes up to the age of 18 months.

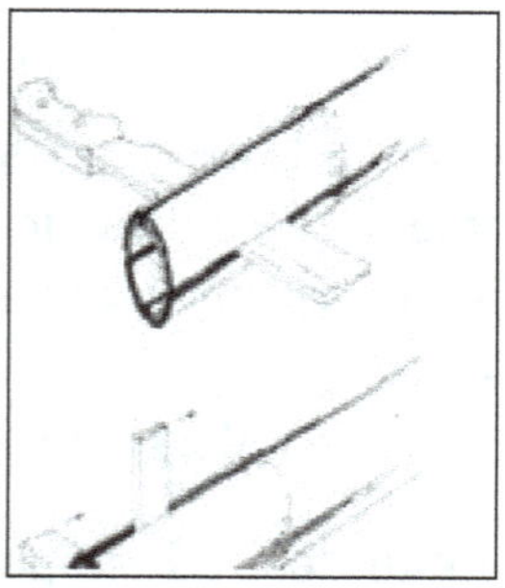

Splitting of a bamboo cane.

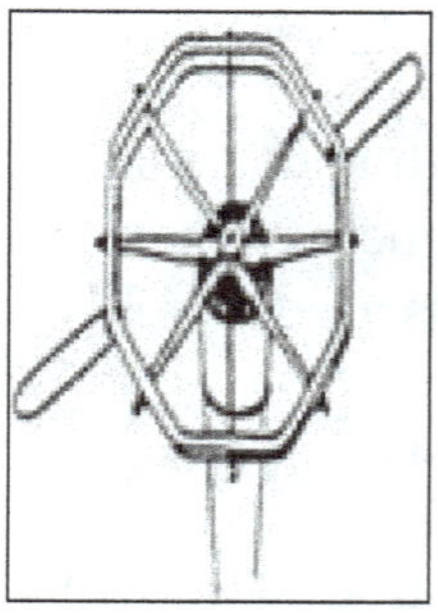

Splitting bamboo with a knife frame.

Shaping of Bamboo

Even though bamboos are naturally circular in form but if they are grown in a box of square shape they acquire a shape as desired.

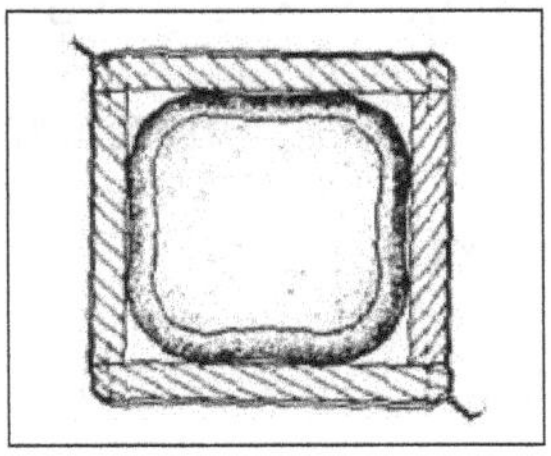

Bending of Bamboo

Bamboos can be bent while they are freshly cut by heating them above the temperature of 150 °C. Bamboo will retain this shape even after cooling and drying off.

Preservation of Bamboo

A thorough treatment of Bamboo is required to protect it against insects and rot before it is put into use. Commonly a mixture of Borax and Boric acid are utilized for this purpose. Another procedure generally employed is to boil cut bamboo to remove the starches that draw insects.

Bamboo as a Building Material

Utilization of Bamboo for construction is achieved by a structural frame technique which is related to same approach applied in usual timber frame design and construction.

In the case bamboo, floor, walls and roof are interconnected and often rely on the other for overall stability. Bamboo has played a vital role in the growth of enterprises and the rural transformation.

Bamboo for Foundations

There is very limited use of bamboo as foundation material because when in contact with moisture laden surface they decay fast. However, this issue can be tackled to quite an extent though proper treatment using appropriate chemicals.

The various types of foundations constructed with bamboo are:

- Bamboo which is in direct contact with ground surface.
- Bamboo fixed to rock or preformed concrete footings.
- Composite bamboo or concrete columns.
- Bamboo piles.

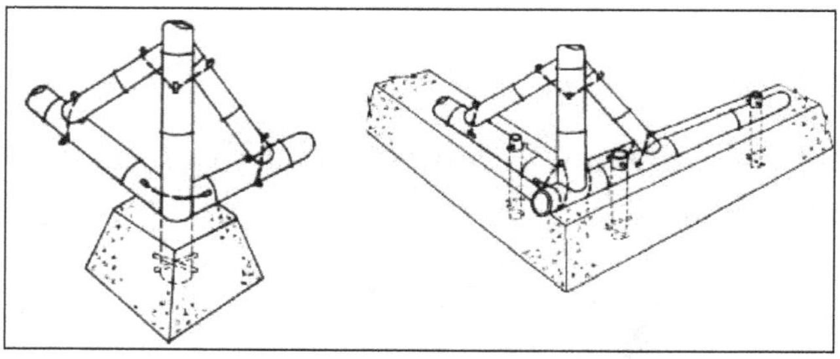

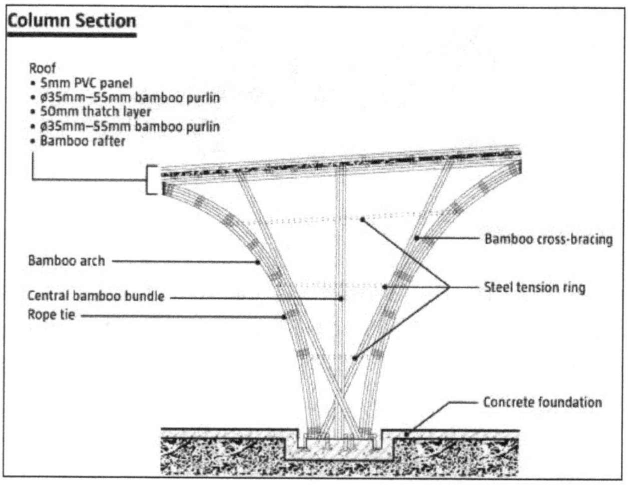

Bamboos are used in various shapes and forms to build foundation. Some of the common shapes of bamboos are:

- Flattened bamboo shape which is acquired by splitting freshly cut bamboo stalks and then rolling and flattening them.
- Bamboo mats as thin as 5-6mm or 10-15mm in size are woven according to design prerequisite. Phenolic resins are used in structures employing bamboo mats.
- Bamboo plastic composite is a pioneering technology in which bamboo fiber as raw material is blended with plastic as the core material. These mats are highly resistant to moisture and structurally more stable.

Walls Construction with Bamboo as a Building Material

Bamboo is extensively used for construction of walls and partitions. Posts and beams are the main elements normally constructed with bamboo provide structural framework for walls. They positioned in a way to be able to withstand forces of nature. An infill is used between framing elements to add strength and stability to the walls.

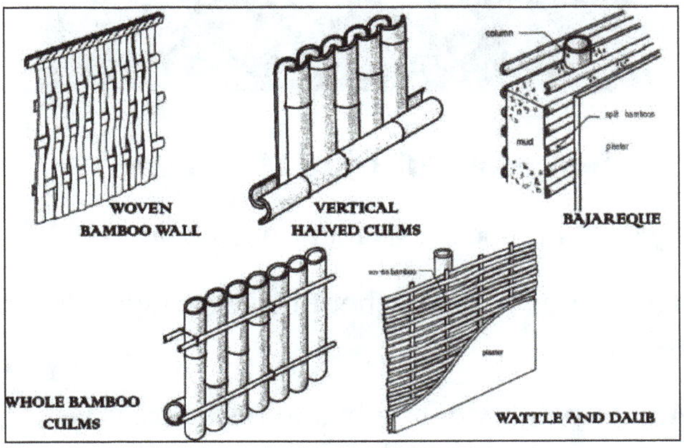

Roofing with Bamboo as a Building Material

Bamboo is one of the best roofing materials and provides ample sturdiness to the structure. It is a proven shield against forces of nature or animals and are considerably light weighted which makes them easy to install. The bamboo roofs encompass purlins, rafters and trusses.

Scaffolding with Bamboo as a Building Material

Due to advantageous properties of bearing heavy load bamboos are considered as one of the highly-endorsed materials for scaffolding even for tall structures.

For the construction of scaffolding, cane extensions are obtained by lashing cane ends using several ropes. The ties are positioned in such a way that forces acting vertically downwards lodge the nodes in the lashing.

This technique has immense significance since the joints can be re-aligned in the right degree.

Advantages of Bamboo as a Building Material

The various advantages of bamboo are as mentioned below:

- Tensile strength: Bamboo has higher tensile strength than steel because its fibers run axially.
- Fire Resistance: Capability of bamboo to resist fire is very high and it can withstand temperature up to 4000 °C. This is due to the presence of high value of silicate acid and water.
- Elasticity: Bamboo is widely preferred in earthquake prone regions due to its elastic features.
- Weight of bamboo: Bamboos due to their low weight are easily displaced or installed making it very easier for transportation and construction.
- Unlike other building materials like cement and asbestos, bamboo poses no danger to health.
- They are cost effective and easy to use.
- They are especially in great demand in earthquake prone areas.

Disadvantages of Bamboo

Bamboos come with their own set of drawbacks such as:

- They require preservation.
- Shrinkage: Bamboo shrinks much greater than any other type of timber especially when it loses water.
- Durability: Bamboo should be sufficiently treated against insect or fungus attack before being utilized for building purposes.
- Jointing: Despite prevalence of various techniques of jointing, structural reliability of bamboo is questionable.

Adobe

Adobe is a building material made from earth and organic materials. Most adobe buildings are similar in appearance to cob and rammed earth buildings. Adobe is among the earliest building materials, and is used throughout the world.

Adobe bricks are rectangular prisms small enough that they can quickly air dry individually without cracking. They can be subsequently assembled, with the application of adobe mud to bond the individual bricks into a structure. There is no standard size, with substantial variations over the years and in different regions. In some areas a popular size measured 8 by 4 by 12 inches (20 cm × 10 cm × 30 cm) weighing about 25 pounds (11 kg); in other contexts the size is 10 by 4 by 14 inches (25 cm × 10 cm × 36 cm) weighing about 35 pounds (16 kg). The maximum sizes can reach up to 100 pounds (45 kg); above this weight it becomes difficult to move the pieces, and it is preferred to ram the mud in situ, resulting in a different typology known as rammed earth.

Strength

In dry climates, adobe structures are extremely durable, and account for some of the oldest existing buildings in the world. Adobe buildings offer significant advantages due to their greater thermal mass, but they are known to be particularly susceptible to earthquake damage if they are not reinforced. Cases where adobe structures were widely damaged during earthquakes include the 1976 Guatemala earthquake, the 2003 Bam earthquake, and the 2010 Chile earthquake.

Distribution

Buildings made of sun-dried earth are common throughout the world (Middle East, Western Asia, North Africa, West Africa, South America, southwestern North America, Spain, and Eastern Europe.) Adobe had been in use by indigenous peoples of the Americas in the Southwestern United States, Mesoamerica, and the Andes for several thousand years. Puebloan peoples built their adobe structures with handsful or basketsful of adobe, until the Spanish introduced them to making bricks. Adobe bricks were used in Spain from the Late Bronze and Iron Ages (eighth century BCE onwards). Its wide use can be attributed to its simplicity of design and manufacture, and economics.

A distinction is sometimes made between the smaller adobes, which are about the size of ordinary baked bricks, and the larger adobines, some of which may be one to two yards (1–2 m) long.

Composition

An adobe brick is a composite material made of earth mixed with water and an organic material such as straw or dung. The soil composition typically contains sand, silt and clay. Straw is useful in binding the brick together and allowing the brick to dry evenly, thereby preventing cracking due to uneven shrinkage rates through the brick. Dung offers the same advantage. The most desirable soil texture for producing the mud of adobe is 15% clay, 10–30% silt, and 55–75% fine sand. Another study states 15–25% clay and the remainder sand and coarser particles up to cobbles 50 to 250 mm (2 to 10 in), with no deleterious effect. Modern adobe is stabilized with either emulsified asphalt or Portland cement up to 10% by weight.

Adobe style.

No more than half the clay content should be expansive clays, with the remainder non-expansive illite or kaolinite. Too much expansive clay results in uneven drying through the brick, resulting in cracking, while too much kaolinite will make a weak brick. Typically the soils of the Southwest United States, where such construction has been widely used, are an adequate composition.

Material Properties

Adobe walls are load bearing, i.e. they carry their own weight into the foundation rather than by another structure, hence the adobe must have sufficient compressive strength. In the United States, most building codes call for a minimum compressive strength of 300 lbf/in^2 (2.07 $newton/mm^2$) for the adobe block. Adobe construction should be designed so as to avoid lateral structural loads that would cause bending loads. The building codes require the building sustain a 1 g lateral acceleration earthquake load. Such acceleration will cause lateral loads on the walls, resulting in shear and bending and inducing tensile stresses. To withstand such loads, the codes typically call for a tensile modulus of rupture strength of at least 50 lbf/in^2 (0.345 $newton/mm^2$) for the finished block.

The Great Mosque of Djenné, Mali, is built in adobe. The struts projecting from the wall serve as decoration, as well as supports for scaffolding during maintenance.

In addition to being an inexpensive material with a small resource cost, adobe can serve as a significant heat reservoir due to the thermal properties inherent in the massive walls typical in adobe construction. In climates typified by hot days and cool nights, the high thermal mass of adobe mediates the high and low temperatures of the day, moderating the temperature of the living space. The massive walls require a large and relatively long input of heat from the sun (radiation) and from the surrounding air (convection) before they warm through to the interior. After the sun sets and the temperature drops, the warm wall will continue to transfer heat to the interior for several hours due to the time-lag effect. Thus, a well-planned adobe wall of the appropriate thickness is very effective at controlling inside temperature through the wide daily fluctuations typical of desert climates, a factor which has contributed to its longevity as a building material.

Thermodynamic material properties have significant variation in the literature. There is an effective R-value for a north facing 10-in wall of Ro=10 hr ft² °F/Btu, which corresponds to thermal conductivity k=10 in x 1 ft/12 in /Ro=0.33 Btu/(hr ft °F) or 0.57 W/(m K) in agreement with the thermal conductivity reported from another source. To determine the total R-value of a wall, scale R_0 by the thickness of the wall in inches. The thermal resistance of adobe is also stated as an R-value for a 10-inch wall R_0=4.1 hr ft² °F/Btu. Another source provides the following properties: conductivity=0.30 Btu/(hr ft °F) or 0.52 W/(m K); specific heat capacity=0.24 Btu/(lb °F) or 1 kJ/(kg K) and density=106 lb/ft³ or 1700 kg/m³, giving heat capacity=25.4 Btu/(ft³ °F) or 1700 kJ/(m³ K). Using the average value of the thermal conductivity as k = 32 Btu/(hr ft °F) or 0.55 W/(m K), the thermal diffusivity is calculated to be 0.013 ft²/h or 3.3×10^{-7} m²/s.

Uses

Poured and Puddled Adobe Walls

Cliff dwellings of poured or puddled adobe (cob) at Cuarenta Casas.

Poured and puddled adobe (puddled clay, piled earth), today called cob, is made by placing soft adobe in layers, rather than by making individual dried bricks or using a form. "Puddle" is a general term for a clay or clay and sand-based material worked into a dense, plastic state. These are the oldest methods of building with adobe in the Americas until holes in the ground were used as forms, and later wooden forms used to make individual bricks were introduced by the Spanish.

Adobe Brick

Bricks made from adobe are usually made by pressing the mud mixture into an open timber frame. In North America, the brick is typically about 25 by 36 cm (10 by 14 in) in size. The mixture is molded into the frame, which is removed after initial setting. After drying for a few hours, the bricks are turned on edge to finish drying. Slow drying in shade reduces cracking.

Adobe bricks near a construction site.

The same mixture, without straw, is used to make mortar and often plaster on interior and exterior walls. Some cultures used lime-based cement for the plaster to protect against rain damage.

Depending on the form into which the mixture is pressed, adobe can encompass nearly any shape or size, provided drying is even and the mixture includes reinforcement for larger bricks. Reinforcement can include manure, straw, cement, rebar or wooden posts. Straw, cement, or manure added to a standard adobe mixture can produce a stronger, more crack-resistant brick. A test is done on the soil content first. To do so, a sample of the soil is mixed into a clear container with some water, creating an almost completely saturated liquid. The container is shaken vigorously for one minute. It is then allowed to settle for a day until the soil has settled into layers. Heavier particles settle out first, sand above, silt above that and very fine clay and organic matter will stay in suspension for days. After the water has cleared, percentages of the various particles can be determined. Fifty to 60 percent sand and 35 to 40 percent clay will yield strong bricks. The Cooperative State Research, Education, and Extension Service at New Mexico State University recommends a mix of not more than 1/3 clay, not less than 1/2 sand and never more than 1/3 silt.

Adobe Wall Construction

The ground supporting an adobe structure should be compressed, as the weight of adobe wall is significant and foundation settling may cause cracking of the wall. Footing depth is to below the ground frost level. The footing and stem wall are commonly 24 and 14 inches thick, respectively. Modern construction codes call for the use of reinforcing steel in the footing and stem wall. Adobe bricks are laid by course. Adobe walls usually never rise above two stories as they are load bearing and adobe has low structural strength. When creating window and door openings, a lintel is placed on top of the opening to support the bricks above. Atop the last courses of brick, bond beams made of heavy wood beams or modern reinforced concrete are laid to provide a horizontal bearing plate for the roof beams and to redistribute lateral earthquake loads to shear walls more able to carry the forces. To protect the interior and exterior adobe walls, finishes such as mud plaster, whitewash or stucco can be applied. These protect the adobe wall from water damage, but need to be reapplied periodically. Alternatively, the walls can be finished with other non-traditional plasters that provide longer protection. Bricks made with stabilized adobe generally do not need protection of plasters.

The earthen plaster removed, exposing the adobe bricks at Fort St. Sebastian.

Adobe Roof

The traditional adobe roof has been constructed using a mixture of soil/clay, water, sand and organic materials. The mixture was then formed and pressed into wood forms, producing rows of dried earth bricks that would then be laid across a support structure of wood and plastered into place with more adobe.

Depending on the materials available, a roof may be assembled using wood or metal beams to create a framework to begin layering adobe bricks. Depending on the thickness of the adobe bricks, the framework has been performed using a steel framing and a layering of a metal fencing or wiring over the framework to allow an even load as masses of adobe are spread across the metal fencing like cob and allowed to air dry accordingly. This method was demonstrated with an adobe blend heavily impregnated with cement to allow even drying and prevent cracking.

The more traditional flat adobe roofs are functional only in dry climates that are not exposed to snow loads. The heaviest wooden beams, called vigas, lie atop the wall. Across the vigas lie smaller members called latillas and upon those brush is then laid. Finally, the adobe layer is applied.

To construct a flat adobe roof, beams of wood were laid to span the building, the ends of which were attached to the tops of the walls. Once the vigas, latillas and brush are laid, adobe bricks are placed. An adobe roof is often laid with bricks slightly larger in width to ensure a greater expanse is covered when placing the bricks onto the roof. Following each individual brick should be a layer of adobe mortar, recommended to be at least 25 mm (1 in) thick to make certain there is ample strength between the brick's edges and also to provide a relative moisture barrier during rain.

Roof design evolved around 1850 in the American Southwest. Three inches of adobe mud was applied on top of the latillas, and then 18 inches of dry adobe dirt applied to the roof. The dirt was contoured into a low slope to a downspout aka a 'canal'. When moisture was applied to the roof the clay particles expanded to create a waterproof membrane. Once a year it was necessary to pull the weeds from the roof and reslope the dirt as needed.

Depending on the materials, adobe roofs can be inherently fire-proof. The construction of a chimney can greatly influence the construction of the roof supports, creating an extra need for care in choosing the materials. The builders can make an adobe chimney by stacking simple adobe bricks in a similar fashion as the surrounding walls.

In 1927, the Uniform Building Code (UBC) was adopted in the United States. Local ordinances, referencing the UBC added requirements to building with adobe. These included: restriction of building height of adobe structures to 1-story, requirements for adobe mix (compressive and shear strength) and new requirements which stated that every building shall be designed to withstand seismic activity, specifically lateral forces. By the 1980s however, seismic related changes in the California Building Code effectively ended solid wall adobe construction in California; however Post-and-Beam adobe and veneers are still being used.

Carbon Fibre

Carbon fibers have high fatigue strength and elastic modulus as compare to glass fibers. Considering service period, studies demonstrate that carbon fiber reinforced polymers have much potential than agamid and glass fibers. They are highly chemically resistant, having high-temperature tolerance with low corrosion resistance and thermal expansion.

Each fiber is size of 510 microns in diameter. To explain how small that is, one micron (um) is 0.00004 inches.

One strand of spider web silk is usually in between 38 microns. Carbon fibers are double as stiff as steel and five times as strong as steel, (per unit of weight). The most imperative elements deciding the physical properties of carbon fiber are a level of carbonization (carbon content, typically over 92% by weight) and an introduction of the layered carbon planes (the strips). Carbon fiber strengthened composite materials are utilized to make airplane and rocket parts, dashing auto bodies, golf club shafts, bike outlines, angling bars, car springs, sailboat poles, and numerous different segments where light weight and high quality are required. Carbon fiber's high quality, light weight, and imperviousness to erosion make it a perfect strengthening material.

Properties of Carbon Fiber

1. Carbon Fiber has high Strength to Weight Ratio (otherwise called particular quality). The strength of a material is the drive per unit range at disappointment, isolated by its thickness. Any material that is solid and light has a good Strength/weight proportion. Materials, for example, Aluminum, titanium, magnesium, carbon and glass fiber, high-quality steel combinations all have a great quality to weight proportions. It is not astonishing that Balsa wood comes in with a high strength to weight proportion.

2. Carbon Fiber is extremely inflexible rigidity or firmness of a material is measured by its Young Modulus and measures how much a material avoids under anxiety. Carbon fiber fortified plastic is more than 4 times stiffer than glass strengthened plastic, just about 20 times more than pine, 2.5 circumstances more noteworthy than aluminum.

3. Carbon fiber is corrosion proof and chemically stable. Regardless of the way that carbon fiber themselves don't deteriorate, epoxy is fragile to light and ought to be secured. Diverse frameworks (whatever the carbon fiber is embedded in) might in like manner be responsive.

4. Carbon fiber is electrically conductive. This element can be valuable and be an aggravation. In boat building, it must be considered similarly as Aluminum conductivity becomes an integral factor. Carbon fiber conductivity can encourage Galvanic Corrosion in fittings. The watchful establishment can decrease this issue. Carbon Fiber clean can gather in a shop and cause starts or short circuits in electrical apparatuses and hardware.

5. Weakness Resistance is great Resistance to Fatigue in Carbon Fiber Composites is great. However, when carbon fiber fizzles it as a rule flops disastrously without much to declare its inescapable break. Harm intractable weakness is viewed as decrease in solidness with bigger quantities of stress cycles; (unless the temperature is high) Test has demonstrated that disappointment is probably not going to be an issue when cyclic burdens correspond with the fiber introduction. Carbon fiber is better than E glass in exhaustion and static quality and in addition solidness. The introduction of the strands and the distinctive fiber layer introduction, impact how a composite will

oppose weakness (as it has on firmness). The sort of powers connected likewise result in various sorts of disappointments. Strain, Compression or Sheer powers all outcomes in particularly extraordinary disappointment comes about. Paper on trial of carbon fiber composites expected for car utilizes. American Institute of Aeronautics and Astronautics, test for materials to be utilized as a part of wind turbines cutting edges.

6. Carbon Fiber has great Tensile Strength. Rigidity or extreme quality is the greatest anxiety that a material can withstand while being extended or pulled before necking, or fizzling. Necking is the point at which the cross segment begins to fundamentally contract. On the off chance that you take a piece of plastic pack, it will extend and at one point will begin getting restricted. This is necking. It is measured in Force per Unit region. Fragile materials, for example, carbon fiber don't generally come up short at a similar anxiety level due to inside defects. They fizzle at little strains. (at the end of the day there is not a ton of bowing or extending before cataclysmic disappointment).

7. Imperviousness to fire/non-flammable depending upon the assembling procedure and the forerunner material, carbon fiber can be very delicate and can be made into or all the more regularly incorporated into defensive attire for fire-fighting. Since carbon fiber is likewise synthetically exceptionally dormant, it can be utilized where there is fire consolidated with destructive specialists. Carbon Fiber Blanket utilized as welding insurance.

8. Thermal conductivity is the measure of warmth transmitted through a unit thickness, toward a way regular to a surface of unit area, in perspective of a unit temperature slant, under tenacious conditions. All things considered it's a measure of how easily warm travels through a material. There are various frameworks of measures relying upon metric or supreme units.

Fibre Reinforced Polymer

Fibre Reinforced Polymer (FRP) composite is defined as a polymer that is reinforced with fibre. It represents a class of materials that fall into a category referred to as composite materials. Composite materials are made by dispersing particles of one or more materials in another material, which forms a continuous network around them.

FRP composites are different from traditional construction materials like Steel and Aluminum. FRP composites are anisotropic whereas Steel and Aluminum are isotropic. Therefore, their properties are directional, meaning that the best mechanical properties are in the direction of the fibre placement.

These materials have a high ratio of strength to density, exceptional corrosion resistance and convenient electrical, magnetic and thermal properties. However, they are brittle and their mechanical properties may be affected by the rate of loading, temperature and environmental conditions.

The primary function of fibre reinforcement is to carry the load along the length of the fiber and to provide strength and stiffness in one direction. It replaces metallic materials in many structural applications where load-carrying capacity is important.

The use of FRP in engineering applications enables engineers to obtain significant achievements in the functionality, safety and economy of construction because of their mechanical properties.

Components of Composite Materials

Fibres

The choice of fibre frequently controls the properties of composite materials. Carbon, Glass, and Aramid are three major types of fibres which are used in construction. The composite is often named by the reinforcing fibre, for instance, CFRP for Carbon Fibre Reinforced Polymer. The most important properties that differ between the fibre types are stiffness and tensile strain.

Glass, Carbon, and Aramid Fibre.

Matrices

The matrix should transfer forces between the fibres and protect the fibres form detrimental effects. Thermosetting resins (thermosets) are almost exclusively used. Vinylester and epoxy are the most common matrices.

Epoxy is mostly favoured above vinylester but is also more costly. Epoxy has a pot life around 30 minutes at 20 degree Celsius but can be changed with different formulations. Epoxies have good strength, bond, creep properties and chemical resistance.

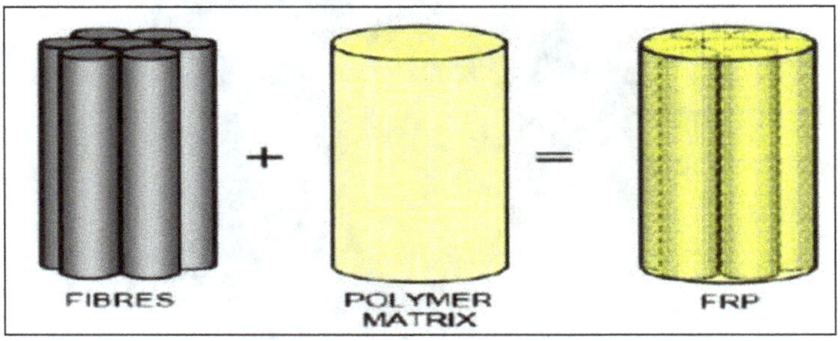

Fibre Plus Matrix produce FRP.

Types of Fibre Reinforced Polymer (FRP)

Glass Fibre Reinforced Polymer (GFRP)

Glass fibres are basically made by mixing silica sand, limestone, folic acid and other minor ingredients. The mix is heated until it melts at about 1260 °C.

The molten glass is then allowed to flow through fine holes in a platinum plate. The glass strands are cooled, gathered and wound. The fibres are drawn to increase directional strength. The fibres are then woven into various forms for use in composites.

Based on an aluminium lime borosilicate composition, glass produced fibres are considered as the predominant reinforcement for polymer matrix composites due to their high electrical insulating properties, low susceptibility to moisture and high mechanical properties.

Glass is generally a good impact resistant fibre but weighs more than carbon or aramid. Glass fibres have excellent characteristics equal to or better than steel in certain forms.

Glass Fibre Reinforced Polymer Bars.

Carbon Fibre Reinforced Polymer (CFRP)

Carbon fibres have a high modulus of elasticity, 200-800 GPa. The ultimate elongation is 0.3-2.5 % where the lower elongation corresponds to the higher stiffness and vice versa.

Carbon fibres do not absorb water and are resistant to many chemical solutions. They withstand fatigue excellently and neither corrodes nor show any creep or relaxation.

Carbon Fibre Reinforced Polymer Bars.

Aramid Fibre Reinforced Polymer (AFRP)

Aramid is the short form for aromatic polyamide. A well-known trademark of aramid fibres is Kevlar but there does exist other brands as well such as Twaron, Technora and SVM.

The modulii of the fibres are 70-200 GPa with an ultimate elongation of 1.5-5% depending on the quality. Aramid has a high fracture energy and is therefore used for helmets and bullet-proof garments.

They are sensitive to elevated temperatures, moisture and ultraviolet radiation and therefore not widely used in civil engineering applications. Finally, Aramid fibres do have problems with relaxation and stress corrosion.

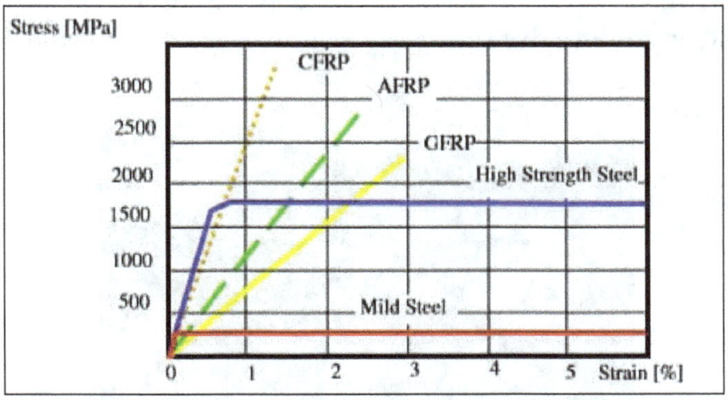

Properties of Different Types of FRP Compared with Steel.

Applications of FRP

- Carbon FRPs are used in prestressed concrete for applications where high resistance to corrosion and electromagnetic transparency of CFRP are important.
- CFRP composites are employed for underwater piping and structural parts of offshore platform. Added to that, FRP declines the risk of fire.
- Carbon fibre reinforced polymers are used to manufacture underwater pipes for great depth because it provides a significantly increased buoyancy (due to its low density) compared to steel.
- The stairways and walkways are also made of composites for weight saving and corrosion resistance.
- It is used in high-performance hybrid structures.
- FRP bars are used as internal reinforcement for concrete structures.
- FRP bars, sheets, and strips are used for strengthening of various structures constructed from concrete, masonry, timber, and even steel.

- FRPs are employed for seismic retrofitting.
- Fibre reinforced polymers are used in the construction of special structures requiring electrical neutrality.
- The high energy absorption of aramid fibre reinforced polymer (AFRP) composites makes them suitable for strengthening engineering structures subjected to dynamic and impact loading.

Mudbrick

A mudbrick or mud-brick is an air-dried brick, made of a mixture of loam, mud, sand and water mixed with a binding material such as rice husks or straw. Though mudbricks are known from 7000 to 6000 BCE, since 4000 BC, bricks have also been fired, to increase their strength and durability.

New, unlaid mudbricks in the Jordan Valley.

In warm regions with very little timber available to fuel a kiln, bricks were generally sun-dried. In some cases, brickmakers extended the life of mud bricks by putting fired bricks on top or covering them with stucco.

Mudbrick was used for the construction of Elamite ziggurats—some of the world's largest and oldest constructions. Choqa Zanbil, a 13th-century BC ziggurat in Iran, is similarly constructed from clay bricks combined with burnt bricks.

The South Asian inhabitants of Mehrgarh constructed and lived in mud-brick houses between 7000–3300 BC. Mud bricks were used at more than 15 reported sites attributed to the 3rd millennium BC in the ancient Indus Valley Civilization. In the Mature Harappan phase fired bricks were used.

Mudbricks were adopted in the Middle East from Indus Valley Cities during the Pre-Pottery Neolithic B period. The Mesopotamians used sun-dried bricks in their city construction; typically these bricks were flat on the bottom and curved on the top, called plano-convex mud bricks. Some were formed in a square mould and rounded so that the middle was thicker than the ends. Some walls had a few courses of fired bricks from their bases up to the splash line to extend the life of the building.

In Minoan Crete, at the Knossos site, there is archaeological evidence that sun-dried bricks were used in the Neolithic period (prior to 3400 BC).

In Ancient Egypt, workers gathered mud from the Nile River and poured it into a pit. Workers then tramped on the mud while straw was added to solidify the mold. The mudbricks were chemically suitable as fertilizer, leading to the destruction of many ancient Egyptian ruins, such as at Edfu. A well-preserved site is Amarna. Mudbrick use increased at the time of Roman influence.

Roofing Materials

The type of roofing material used will depend on the style of home you are building, the slope of the roof, and local building restrictions. The following is a list of the roofing materials that are most commonly used in the construction of today's homes.

Asphalt Shingles

This is the most common type of roofing material used today. This type of shingle is divided into two different categories, organic or fiberglass (non-organic). Organic shingles are manufactured with a cellulose fiber base made from recycled paper and wood fibers, which is then saturated with asphalt. Then a mineral coating is applied to resist weathering. Fiberglass shingles are produced in a similar fashion but the core is made up of fiberglass. This tends to provide more flexibility and added strength over the organic. Dimensional shingles are a variation of organic or fiberglass shingles. They are similar to the standard organic or fiberglass shingle, however, the cores are much thicker. The additional layers of material can be sculpted to provide different shadow lines to give the roof a more custom look.

Asphalt shingles are manufactured in a wide variety of colors and are rated by their projected lifespan. This is typically 20-30 years for the standard organic or fiberglass, while dimensional shingles can have a life expectancy up to 40 years. The majority of manufacturer's will only warranty their materials for the specified life spans if their certified roofers install them.

Wood Shingles

The majority of wood shingles comes from Western red cedar. The reason this type of wood is used most often is its resistance to decay. The wood shingles are classified and sold as either number 1, 2, or 3. 1 is what is used and recommended for roofing because it is cut from knot free heartwood. 2 is generally used for siding and comes from less resistant sapwood. This type of shingle comes in a few different sizes, 16 inch, 18 inch, and 24 inch, and is tapered.

Wood Shakes

Like shingles this type of roofing material is taken from cedar. The difference is that shakes are either sawn again or split by hand. A resawn shake has one side that is left with a natural irregular look and one side that is cut thinner to give the tapered shape. Hand split shakes are tapered as well but typically have more natural, textured look. Shakes are separated by weight, which are classified as either heavy or medium and are sold in lengths of 18 inch or 24 inch.

Clay and Concrete Shingles

This style of roofing is usually associated with the Spanish-mission tiles. The most common shape of this style of roofing material is the traditional barrel shape, but you will also find clay and concrete tile that is flat or tapered to give a wood shake or slate appearance. The life span can be between 50-100 years, but this type of material performs best in warm dryer climates. Tile tends to absorb water, so this can be a problem in moist or cold climates where the tile could be damaged due to the freeze/thaw cycle. Concrete and clay tiles are heavy so it is often necessary to increase roof support for the added weight.

Metal

Roofs constructed using metal shingles have been around for years. Today's advances have made it available in many different colors, shapes and styles. Metal shingles will typically be interlocking and can be made up of aluminum, copper or stainless steel. This type of roofing is lightweight and doesn't crack, split, rot or burn and will typically last over 50 years. Metal can also be used on many different styles and both steep and flat roofs.

Slate

This is one of the oldest roofing materials found U.S. and is mainly used in the northeast. Slate is a hard rock that can be split into thin slabs, which can vary between 1/4 inch to 1 inch. Slate is extremely durable and can last 100 years or more. This type of roofing material is very heavy so it requires the roof be built specially to handle the added weight, and can be difficult to install properly, so it can be a problem finding a qualified roofer.

References

- Stainless-steel-reasons-why-its-used-widely-construction-industry: srmbsteel.com, Retrieved 04 April 2019
- How-is-stainless-steel-used-in-the: bsstainless.com, Retrieved 11 June 2019
- Concrete-building-material, technology: britannica.com, Retrieved 27 August 2019
- Reinforced-concrete, technology: britannica.com, Retrieved 18 July 2019
- Aluminium-building-construction-industries: laminazionesottile.com, Retrieved 31 March 2019
- Composite-material, technology: britannica.com, Retrieved 24 January 2019

4

Structural Stability and Failure

Structural stability aims at ensuring the safety of structures against failures in engineering designs and plans. The basic types of structural failure include compressive, tensile, bending and buckling. This chapter discusses about the aspects of structural stability and failure in detail.

Structural Stability

Stability represents a fundamental problem in solid mechanics, which must be mastered to ensure the safety of structures against collapse. The theory of stability is of crucial importance for structural engineering, aerospace engineering, nuclear engineering, offshore, ocean and arctic engineering. It plays an important role in certain problems of space structures, geotechnical structures, geophysics and materials science.

Stability of Columns, Frames and Arches

The concept of a critical load of an elastic structure at which the equilibrium bifurcates was introduced by Euler who also provided the solutions of critical loads of columns with various end restraints. Experiments, however, could not verify the calculated critical loads. This fact was explained by Young, who realized that imperfections such as initial curvature, initial bending moments or load eccentricity play an important role and derived a formula for what is known today as the magnification factor for deflections and bending moments in columns due to axial load. Kirchho extended the theory to geometrically non-linear large deflections and provided an elegant solution of the deflection curve, called the elastic, in terms of elliptic integrals. The effect of shear, which is manifest in columns with low effective shear stiffness, was clarified by Engesser (ignorance of his solution decades later was unfortunately the prime cause for the collapse of the Quebec Bridge in 1907, precipitated by buckling of one latticed diagonal of a truss having insufficient shear stiffness).

The flexibility method of analysis of frames was extended to critical load analysis by formulating the dependence of the flexibility matrix of the column on its axial force, and the same was soon done for the stiffness matrix. The flexibility method, applied to the primary statically determinate structure of a redundant frame, can also be used but can be misleading if there are many statically indeterminate internal forces because the flexibility matrix of the primary structure, unlike the stiffness matrix of the original structure, can and typically does, lose positive definitiveness before the first critical load.

The matrix stiffness method in the form of finite elements of beams has been proven more suitable for computer analysis and has made the calculation of the critical loads of elastic frames a routine problem. For large regular frames, the critical loads can be obtained analytically by methods of difference calculus. Even simpler analytical solutions can be obtained by approximating the regular frame with a micro polar continuum.

A considerably more difficult problem is the buckling of slender high arches or rings. Boussinesq's initial solution of a two-hinge arch was later corrected by Hurlbrink, but a good understanding of arbitrary statically indeterminate arches was not reached until the 1970's.

Dynamic Instabilities and Chaos

A structure can lose stability while under accelerated motion. The treatment of dynamic instabilities necessitates a general stability definition which was contributed by Liapunov. Roughly, it states that the motion of a structure is stable if any possible small change in the initial conditions can lead only to a small change in the response. This is important for non-conservative loads, for example, those produced by wind and generally by fluids or by jet propulsion. Solutions of instabilities of columns under various idealized non-conservative loads such as the follower forces occupied mechanicians in the middle of the century.

Dynamic instability, also called flutter, is an important consideration for aircraft wings, suspension bridges, tall chimneys, guyed masts and other structures. Another type of instability, important for foundations of rotating machinery as well as bridge columns, is the parametric resonance, engendered by the fact that the axial displacement of a column has double the frequency of its lateral vibrations, permitting them to resonate with a load of that frequency. An idealized form of the problem leads to Mathieu differential equation, which was solved approximately by Rayleigh. An important special case are conservative systems, for which a theorem due to Lagrange and Dirichlet states that the system is stable if its potential energy is positive definite. This theorem makes it possible to forgo dynamic analysis and reduce the stability problem to an investigation of the shape (and topology) of the potential energy surface as a function of the generalized displacements of the structure. Only limited success has been obtained in a search for functions similar to potential energy, called Liapunov functions, which would decide the stability of non-conservative systems.

The Coriolis (gyroscopic) force, even though it does no work, was found to be the reason for stability of shafts rotating at supercritical speeds. An interesting phenomenon is that non-conservative systems such as rotating machinery stabilized by gyroscopic forces, fluid conveying pipes, aircraft wings and structures under follower forces can be destabilized by damping.

Energy Methods, Post-critical Behavior and Catastrophe Theory

The Lagrange-Dirichlet theorem reduces stability analysis of conservative systems to a check of the positive definiteness of the tangential stiffness matrix of the structure. As a consequence of Liapunov's theorem, the critical loads can be determined from the stiffness matrix of the linearized system, for which the potential energy is quadratic.

The post-critical behavior is characterized by the higher than quadratic terms of the potential energy as a function of generalized displacements. The basic types of post-critical behavior can be classified as stable symmetric (which is always imperfection insensitive), unstable symmetric, and asymmetric (which are both imperfection sensitive, the latter more than the former). For all systems, the initial post-critical behavior is described by Koiter's power laws, a celebrated result of stability theory according to which, for every elastic structure, an imperfection causes a reduction of the maximum load proportional to either the 2/3 or the 1/2 power of the imperfection magnitude, for all elastic systems. Moderate reductions due to imperfections occur in some types of elastic frames, but in cylindrical shells subjected to axial compression or bending and in spherical domes the maximum load reduction due to inevitable imperfections is major, down to about 1/8 to 1/3 of the critical load.

Another important type of instability of elastic systems is the snap through. It occurs in nonlinear systems in which bifurcation with symmetric deflections does not exist, e.g., in flat arches or shallow shells.

The topology of the potential energy surface near the critical load can give rise to very intricate post critical behavior. Similar behavior occurs in many problems of physics and other sciences. It has recently been intensely studied in the theory of catastrophes. A famous result is Thom's proof that for systems with no more than two generalized displacements and no more than four control parameters involving the loads and imperfection magnitudes, there exist no more than seven fundamental catastrophes, called the fold, cusp, swallowtail, butterfly, hyperbolic umbilic, elliptic umbilic, and parabolic umbilic.

The potential energy is also useful as the basis of approximate solutions of critical loads and post critical behavior. A fundamental role is played by Rayleigh's quotient whose evaluation on the basis of an approximate deflection curve or surface is known to yield an upper bound on the exact critical load. When a column or beam structure is statically determinate, a closer upper bound is obtained from the Timoshenko quotient, which is nothing but the Rayleigh quotient for a reduced-order (second-order) differential equation for beam-column. Minimizing the Rayleigh quotient calculated from the deflection surface expressed as a linear combination of a complete system of chosen linearly independent functions is equivalent to the Ritz direct variational method. The minimization approaches the critical load from above.

In view of structural safety, methods providing lower bounds would be preferable. However, this problem is much more difficult, and the available bounds are usually not close.

The potential energy expression is also useful for deriving the differential equation of the problem and the boundary conditions via the calculus of variations.

Thin-Wall Beams, Plates and Shells

Long thin-wall beams, such as metallic cold-formed profiles and steel or concrete girders for bridges or buildings represent long shells which can be approximately treated by a semi-variational approach (Kantorovich variational method), in which the basic deformation modes of the cross-section are judiciously selected so that the potential energy, as well as the corresponding differential equation resulting by variational calculus, be one-dimensional. For open cross-sections, the most important deformation mode is the warping of the cross-section, with the bimoment being the associated force variable. The problem of warping torsion is amenable to simple formulas for lateral buckling and axial torsional buckling of beam columns. For more complicated deformation modes characterizing box cross-sections, systems of ordinary differential equations have been solved numerically, both for linear analysis of the critical load and for large nonlinear post-critical deflections.

The critical loads of elastic rectangular or circular plates are easily solved by series expansions. An interesting point is that among the many critical loads of plates (as well as shells), the lowest one often does not correspond to the longest wavelength of the deflection profiles. Studies of postcritical behavior, beginning with von Kármán and Föppl, have shown that plates are generally not imperfection sensitive and, in fact, possess normally a huge postcritical reserve, which is, however, mobilized only at very large deflections. The limit capacity in postcritical deflections is reached by plasticization of the plate, which develops ridge-shaped buckles causing the plate to act approximately as a truss. Using such a truss analogy, simple formulas have been developed for the maximum loads of rectangular plates, with the remarkable property that the maximum distributed load is independent of the plate dimensions.

The theory of shell buckling has had a fascinating history with a long gestation. After the critical loads of pressurized spherical and axially compressed cylindrical shells were solved early in the century, experiments showed failure loads that were about 3 to 8 times smaller.

Usually the rise of the buckles on the shell surface can be considered as shallow. This is assumed in the shallow shell theory, in which the problem of critical load of a cylindrical shell can be reduced to one 8th order partial differential equation for the deflection. For general shells, the problem leads to a system of eight first-order partial differential equations, known as the Donnell-Mushtari-Vlasov theory. In sandwich plates and shells, interaction of global and local buckling is important.

Buckling of Elasto-Plastic Structures

In 1889, Engesser suggested that the critical load of an inelastic column is obtained by simply replacing the elastic modulus with the tangent modulus for loading. But in 1895 he reversed himself by proposing that a certain geometry-dependent weighted average of the moduli for loading and unloading, called the reduced modulus, should be used. This theory was later refined and extended by von Kármán. After blatant disagreements with measurements on aluminum alloys were detected in aeronautical industry, Shanley, in an epoch-making paper, showed that Engesser's original proposal, namely the initial tangent modulus value, should be used because the column does not buckle at constant load, but at increasing load.

Shanley's theory, which was generalized by Hill, is today generally accepted for calculating the first bifurcation of an elastoplastic structure. The fact that a structure must buckle at its first bifurcation load was later established by analysis of imperfections, and still later much more easily on the basis of entropy increment calculation. A salient feature of elastoplastic buckling is that the structure is not at the stability limit at the bifurcation state and that the deflected postbifurcation states are stable.

The distinction between the critical loads of Engesser's reduced modulus theory and Shanley's tangent modulus theory is small for materials such as mild steel, which reach a horizontal yield plateau abruptly. However, hot-rolled steel profiles indicated a large difference between the tangent modulus load and the reduced modulus load. This fact had remained puzzling until it was discovered that the reason consists in large thermal stresses locked in after cooling, which cause that the diagram of the axial force versus shortening of the column is smoothly curved, without a sudden transition to a yield plateau.

Measurement of bifurcation loads further provided a surprising result with important consequences for the theory of plastic constitutive equation. Tests of torsional buckling of cruciform columns, in which the critical load depends on the initial tangent modulus for shear, revealed that Hencky's simple deformation theory, criticized in other respects, gives correct results while the incremental plasticity theories based on a single loading surface (e.g. von Mises or Tresca) give values much too high. Very complicated behavior is observed in reinforced concrete columns, in which inelastic deformations are combined with tensile cracking and bond slip.

Large plastic deflections of columns need to be understood for predicting the energy absorption capability of the impact, blast, or earthquake. For very large deflections of very slender columns, one can assume formation of plastic yield hinges, which greatly simplifies the calculations. A difficult problem, attacked by finite-strain finite element solutions, has been the localization of plastic strain such as necking in tensioned bars. Plastic localization instabilities are important for the bursting of pipes and other shells due to internal pressure, for bending failure of tubes due to ovalization of the cross-section, and for postcritical reserves in plates and thin-wall girders.

Thermodynamic Analysis of Structural Stability

Although stability of an inelastic structure can be decided by analyzing the effects of all possible imperfections, it is much simpler and more general to use a thermodynamic approach. Since an inelastic structure is normally far from a state of thermodynamic equilibrium at which all the dissipative processes would come to a standstill, the use of irreversible thermodynamics would in principle be necessary.

The classical thermodynamics, which deals only with states infinitely close to thermodynamic equilibrium (and is much simpler), can nevertheless be used by introducing the hypothesis of a tangentially equivalent inelastic structure. The existence of such a structure is of course tacitly implied in finite element programs in which the loading increments are analyzed on the basis of tangential stiffness. Because various combinations of loading and unloading are possible, there is generally a number of tangentially equivalent elastic structures to consider.

Having reduced the problem to elastic structures, one can introduce the incremental internally produced entropy of the structure-load system as well as thermodynamic potentials such as the incremental total energy, Gibbs' or Helmholtz's free energy or enthalpy, which represent the thermodynamic potentials under adiabatic and isothermal conditions, and with either the generalized displacements or the associated forces as the variables. Simple thermodynamic criteria for stability of inelastic structures in the pre-peak and post-peak deflections and for snapback behavior under various types of load control, displacement control, and mixed control have been established. Generally it appears that stability is decided by the positive definiteness of the second variation of these potentials, which represents the second-order work based on the tangential stiffness matrix and is equivalent to the negative of the entropy increment of the structure-load system.

Thermodynamic analysis makes it also possible to determine which branch is followed by the structure after a bifurcation. To this end, one may consider a deviation from equilibrium at constant values of independent variables and changing controls (loads), and subsequent approach at constant controls to a new equilibrium on one or another branch of the post-bifurcation equilibrium path. The path that is followed is that for which the second-order increment of entropy on approach to the new equilibrium state is maximized.

The bifurcation state itself is indicated by singularity of the tangential stiffness matrix. The tangential stiffness matrix that decides the first bifurcation is determined under the assumption that the tangential modulus for loading applies at all points of the structure. When the lowest eigenvalue of this matrix becomes negative, the bifurcation point has been passed. But the structure is not necessarily unstable, and for this reason, a bifurcation state can be missed in a finite element program if the tangential stiffness matrix is not calculated and checked.

More complicated behavior is encountered when phenomena such as friction or damage cause the tangential stiffness matrix to be non-symmetric. Singularity of that matrix decides bifurcation; however, stability is decided strictly by the symmetric part of the stiffness matrix, whose eigenvalue is known to be smaller or equal to the lowest eigenvalue of the non-symmetric matrix.

Stability analysis of load cycles in an elasto-plastic material provides important restrictions for the constitutive laws. It turns out that, in frictional materials, a non-associated flow rule does not necessarily cause instability. This can be analyzed on the basis of the so-called frictionally-blocked second-order energy density.

Damage Localization Instabilities

Localization of damage is a favored mechanism of failure of inelastic structures whose material exhibits strain-softening damage. Such damage is described by stress-strain relations that exhibit a postpeak descent of stress at increasing strain, and in general a loss of positive definiteness of the tangential stiffness matrix of the material. Stress-strain relations of this kind have been used empirically for concrete since the 1950's. However, mechanicians who understood the implications for stability regarded all studies of strain softening damage for several decades with contempt until it was realized that the concept of strain softening can be put on a sound basis by introducing a characteristic length of the material. Curiously, the continuum damage mechanics escaped such contempt despite the absence of material length, perhaps because its strain-softening features were hidden in a separate damage variable while the `true' stress exhibited no strain softening. In absence of a characteristic length, the material cannot propagate waves (loading waves, not unloading ones), the dynamic boundary value problem becomes ill-posed, and the partial differential equation changes its type from hyperbolic to elliptic. There were intense polemics on these questions until about 1985.

Strictly mathematically, the concept of strain softening does make sense even without the material length. Unique exact solutions to some wave propagation problems have been given. However, they exhibit physically unacceptable features. The dynamic problem is ill-posed, and as soon as postpeak strain softening is triggered, the damage instantly localizes into a zone of measure 0 (a point, a line, a surface, with zero volume). Thus, the structure is indicated to fail with zero energy dissipation.

The onset of strain softening, however, can generally be analyzed without introducing the material length. Such analysis was pioneered by Rudnicki and Rice and Rice, who solved the effect of the geometrically nonlinear features of finite strain on localization of (non-softening) plastic strain into an infinite layer of arbitrary thickness within an infinite body. A similar approach to the onset of localization (or bifurcation) has later been pursued for softening materials (in which case the finite strain features become unimportant if the softening is steep). The bifurcation is indicated by the singularity of the so-called acoustic tensor of the material when the orientation of the localization layer is fixed, or the singularity of the tangential stiffness tensor of the material when the orientation is arbitrary. Localizations can also be triggered by a lack of normality of plastic flow in the case of non-associated flow rule. In the case of infinite body, the bifurcation condition represents also the stability limit, but for a finite body the thickness of the localization layer becomes important for stability, and stability is lost later. This was shown first for a strain softening bar and later for a layer of finite thickness in a finite body. The conditions for bifurcation and for the loss of stability have also been analytically formulated for ellipsoidal localization domains, on the basis of Eshelby's theorem. Dynamic bifurcations with localization have been shown to occur under seismic loading in concrete structures.

Strain-softening constitutive relations as an approximation to distributed cracking in reinforced concrete structures have been introduced into finite element analysis by Rashid, Murray and others. Such computational approaches have, however, been shown unobjective with respect to the choice of the mesh size and geometry (exhibiting incorrect convergence on mesh refinement).

The realization that objectivity of finite element calculations necessitates introducing a characteristic length of the material led first to the formulation of the crack band model, and then to transplanting the non-local concept from elasticity into the analysis of strain-softening damage. The crack band model, and to a lesser extent the non-local damage model, have undergone many refinements and have found wide practical applications, especially in the analysis of concrete structures and geotechnical excavations. However, the nonlocal model often requires inconveniently small finite elements, and to deal with such cases finite elements with embedded discontinuities (either an embedded band with strain discontinuity or an embedded line with displacement discontinuity) were introduced. As an approximation to the nonlocal averaging integral, which seems to have some computational advantages, a second-gradient model for strain softening has been proposed. Its effective form solves the nonlocal strain from the local strain from a system of separate Helmholtz partial differential equation. Intricate slip localization instabilities have been found in velocity dependent friction.

Stability Problems of Fracture Propagation

Fracture mechanics presents numerous stability problems, especially when different crack tips interact. In the case of a single crack tip, the limit of stability of crack propagation is reached when the curve of the energy release rate at constant load versus crack length becomes tangent to the R-curve of the material. For some fracture geometries, and for a sufficiently large structure size, crack propagation can lead to snapback instability of the structure. In the case of the cohesive crack model, the stability limit is given in terms of a certain integral equation over the length of the cohesive zone. Simultaneous growth of many cracks typically leads to bifurcations as well as stability loss, which can be analyzed on the basis of the tangential stiffness matrix expressed in terms of the partial derivatives of the stress intensity factor of each crack tip with respect to the length of every crack in the structure.

Often it is found that, in a homogeneous body, simultaneous propagation of several crack tips does not represent the stable path of the system. Rather, the fracture growth localizes into a single crack and the other cracks stop growing or start unloading. Important examples of such behavior are parallel cracks caused by cooling or drying shrinkage in porous materials. The result is that when the parallel cracks reach a certain depth, every other crack stops growing and the intermediate ones propagate further until again every other crack stops growing, etc. In this manner, stability considerations govern the spacing of open parallel cracks. The problem is of interest for the pavements of runways and highways, in geology for the interpretation of drying cracks in mud or cooling cracks in ancient lava flows, etc.

Finite Strain aspects of Stability in Three Dimensions

To determine the tangential stiffness matrix and the critical state, the potential energy must be expressed correctly up to the quadratic terms in displacements. This means that the finite strain tensor must be expressed correctly up to the second-order terms. However, there are many types of finite strain measure which have different second-order terms (e.g. Green's Lagragian, Biot's, Hencky's, etc.). For each of them, the incremental equations of equilibrium, the tangential stiffness matrix and the critical loads are given by differently looking expressions. This fact has caused a long lasting confusion in the stability theory of three-dimensional bodies. A number of different theories which apparently were giving very different results were proposed by Southwell, Biezeno and Hencky, Trefftz, Pearson, Hill, Haringx, Neuber and others.

This zoo of formulations led to controversies. It transpired, however, that all these theories are equivalent because the constitutive relations, and thus the tangential stiffness moduli of the material, are different, with different moduli values for theories associated with different quadratic components of the finite strain tensor. Simple relations between the tangential elastic moduli for different theories have been established. In this manner, it was for example shown that Engesser's and Haringx's formulae for shear buckling of columns are not in contradiction but in fact identical (although the shear modulus must of course be measured for each formula in a different manner). The aforementioned problem does not arise for beams, plates and shells without shear because in those cases the second-order part of the finite strain tensor depends only on the rotations of the cross-sections, in which there is no ambiguity.

Finite strain theory is needed to solve three-dimensional internal buckling of solids, surface buckling, buckling of thick columns and thick tubes, bulging of compressed bars, etc. The critical loads for these buckling modes are generally of the same order of magnitude as the tangential stiffness moduli of the material. This means that such instabilities can occur only in materials that exhibit a high degree of orthotropy, with a low shear stiffness, a condition that can arise as a result of material damage (especially oriented cracking), fibrous microstructure or latticed microstructure (e.g. built-up latticed columns). In compressed fibre composites, the three-dimensional instability gives rise to propagation of kink bands, which control compression strength.

Buckling of Viscoelastic and Viscoplastic Structures

Time dependence of material behavior, that is viscoelasticity or viscoplasticity, may lead to instabilities that develop not suddenly but over a long period of time. If the material is linearly viscoelastic and is a solid, a slender structure possesses a long-time critical load, which is the load that must be reached or surpassed for an infinitesimal imperfection to lead to a finite deflection in infinite time. The long-time buckling problem may be solved by replacing the elastic moduli in the elastic formulation by the corresponding viscoelastic operator, which may be of rate type or integral type. The long-time critical load, however, need not be important for design. Important is the time to reach, for given imperfections, the maximum tolerable deflection or stress due to buckling. This time must exceed the design life time.

When the material is viscoplastic, there exists, in contrast to viscoelastic materials, a finite critical time at which the deflection triggered by an infinitely small imperfection becomes finite. The critical time for viscoplastic buckling of metals controls the temperature to which various mechanical parts of heat engines can be exposed, determine the demands for insulation of steel structures against fire, etc. Considerable complications arise in the analysis of long-time buckling of concrete structures, due to chemically induced long-time aging of the material as well as distributed cracking and its gradual spread with time. Long-time buckling is particularly important for thin shell concrete roofs.

Stability of Concrete Structures

A stable structure shall remain stable for any imaginable system of loads. That is why, load types and point of applications are not taken into considerations while a decision is made about whether a structure is stable or not.

Furthermore, when a structure incapable of meeting the above requirement, geometry of structural component will change under compression, resulting in loss of ability to resist loading and it would be unstable. Instability can lead to catastrophic failure that must be accounted in design.

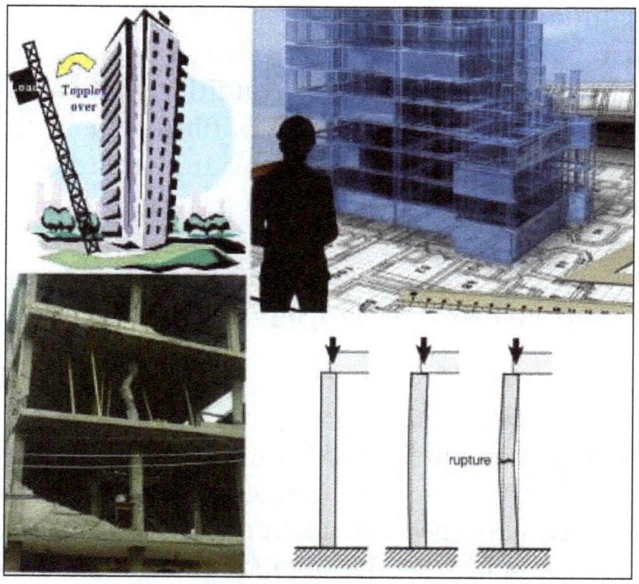

Basic concepts of stability of structure.

Stability Criteria

1. If upon releasing the structure from its virtually displaced state the structure returns to its previous configuration, then the structure is in stable equilibrium.

Moreover, the structure is in stable equilibrium when small perturbations do not cause large movements like a mechanism. Structure vibrates about it equilibrium position. A mechanism cannot resist loads and is of no use to the civil engineer.

2. However, if the structure does not return to its original state following the release of the virtual displacements, the condition is either neutral equilibrium or unstable equilibrium.

Therefore, structure is in unstable equilibrium when small perturbations produce large movements, and the structure never returns to its original equilibrium position.

Finally, structure is in neutral equilibrium when it is not clear whether it is in stable or unstable equilibrium. Small perturbation causes large movements, but the structure can be brought back to its original equilibrium position with no work.

The concept of the stability of various forms of equilibrium of a compressed bar is explained by using the equilibrium of a ball as shown below:

Stable Equilibrium

If a force displaces the ball slightly from its original position of equilibrium, it will return to that position upon the removal of the disturbing force. A body that behaves in this manner is in a state of stable equilibrium.

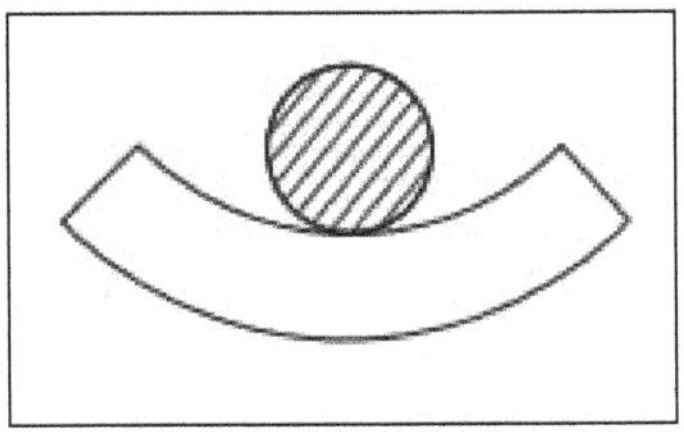

Stable equilibrium.

Unstable Equilibrium

When a disturbing force displaces the ball slightly from its position of equilibrium, it does not return but continues to move down from the original equilibrium position. In this case, the equilibrium of the ball is unstable.

Neutral Equilibrium

If disturbing force displaces the ball slightly, it neither returns to its original equilibrium position nor continues to move away upon removal of the disturbing force. This type of equilibrium termed neutral equilibrium.

If the equilibrium is neutral, there is no change in energy during a displacement in the conservative force system.

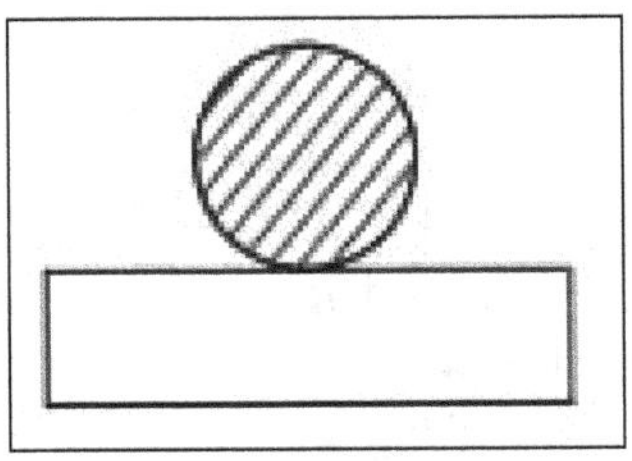

Neutral equilibrium.

Buckling vs. Stability

- Change in geometry of structure under compression – That results in its ability to resist loads – called instability.
- Not true – This termed buckling.
- Buckling is a phenomenon that can occur for structures under compressive loads:
    - The structure deforms and is in stable equilibrium.
    - As the load increases, the structure suddenly changes to deformation at some critical load P_{cr}.
    - The structure buckles.

- Buckling and Stability:
  - The question is – Is the equilibrium in stable or unstable?
  - Usually, after buckling is either neutral or unstable equilibrium.

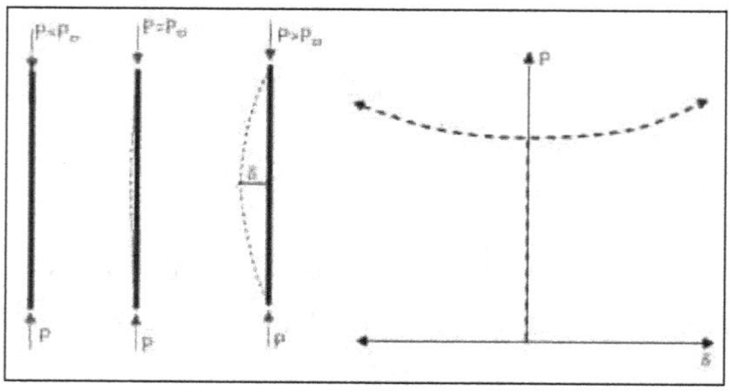

Phases of column buckling.

Types of Instability

Structure subjected to compressive forces can undergo:

- Buckling – bifurcation of equilibrium from deformation:
  - Bifurcation buckling occurs for columns, beams, and symmetric frames under gravity loads only.
- Failure due to instability of equilibrium due to large deformations or material inelasticity:
  - Elastic instability occurs for beam-columns, and frames subjected to gravity and lateral loads.
  - Inelastic instability can occur for all members and the frame.

Bifurcation Buckling

- Member or structure subjected to loads. As the load increases, it reaches a critical value where:
  - The deformation changes suddenly from.
  - The equilibrium load-deformation path bifurcates.
- Critical buckling load when the load-deformation path bifurcates:
  - Primary load-deformation path before buckling.
  - Secondary load-deformation path post buckling
  - Is the post-buckling path stable or unstable.

Symmetric Bifurcation

- Post-buckling load-deform. paths are symmetric about load axis:
    - If the load capacity increases after buckling then stable symmetric bifurcation.
    - If the load capacity decreases after buckling then unstable symmetric bifurcation.

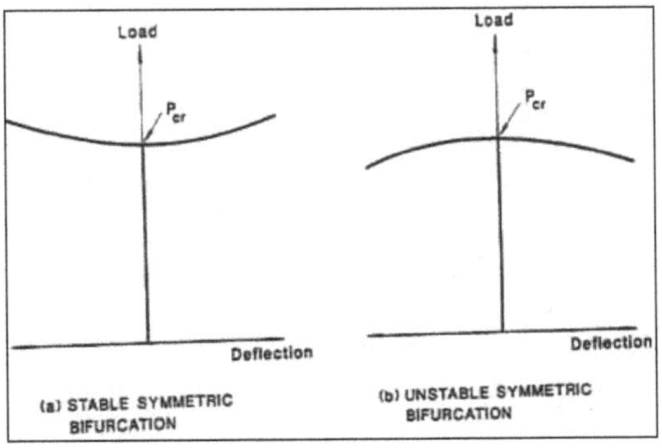

Stable symmetric bifurcation.

Asymmetric Bifurcation

Post-buckling behavior that is asymmetric about load axis.

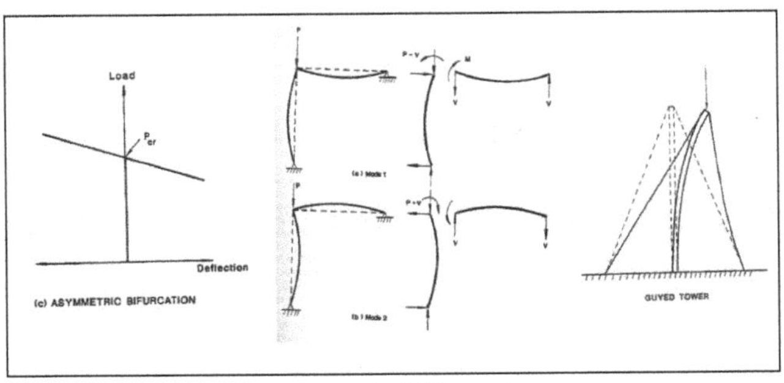

Asymmetric bifurcation.

Instability Failure

- There is no bifurcation of the load-deformation path. The deformation stays in state-1 throughout.
- The structure stiffness decreases as the loads increases. The change is stiffness is due to large deformations and / or material inelasticity:
    - The structure stiffness decreases to zero and becomes negative.
    - The load capacity is reached when the stiffness becomes zero.

- Neutral equilibrium when stiffness becomes zero and unstable equilibrium when stiffness is negative.
- Structural stability failure – when stiffness becomes negative.

Failure of Beam-Columns

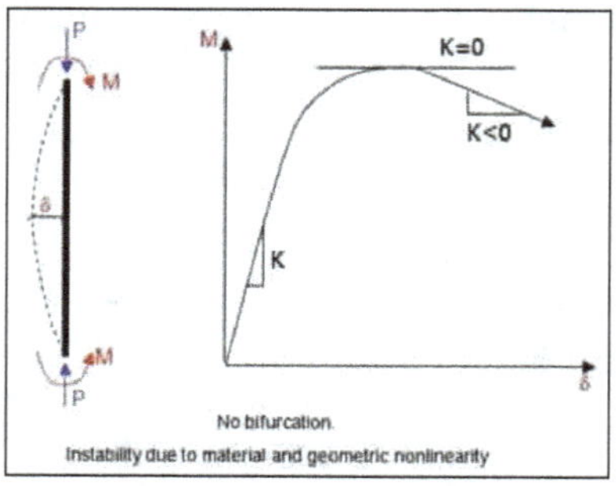

Instability due to material and geometric nonlinearity.

Snap-Through Buckling

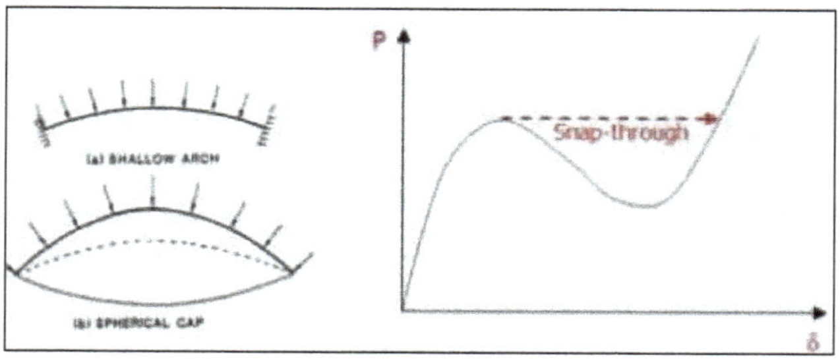

Snap through buckling.

Shell Buckling Failure – Very Sensitive to Imperfections

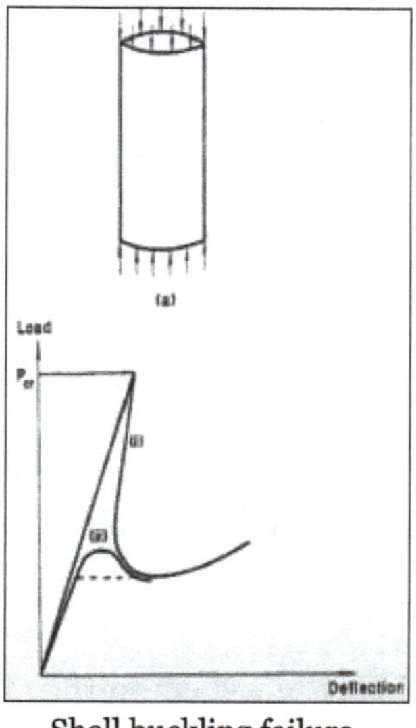

Shell buckling failure.

Structural Failure of Buildings

Defective construction that causes failure may be due to numerous reasons that may not be easy to predict before or during the construction. The major causes of structural failure are defective designs that have not determined the actual loading conditions on the structural elements. Inferior construction materials may also be the cause since the loads are calculated for materials of specific characteristics. Structure may fail even if the design is satisfactory, but the materials are not able to withstand the loads. Employment of unskilled labor on construction work is another reason for structural failures. Therefore, it is important that the owners, designers, and builders are fully conscious of the reasons of failure, and undertake all preventive measures.

Design Deliberations

Construction imperfection in design and manufacturing can be extremely expensive to settle. Architectural design and construction defects cause a structure to be improper for its proposed intent. Correct structural design is significant for all buildings, but exceptionally essential for tall buildings. Even a slight probability of failure is not acceptable since the results can be disastrous for human life and property. Therefore, civil engineers are required to be exceptionally careful and methodical in ensuring an appropriate building design that can sustain the applied loads. All failure modes need to be examined by using modern software on the subject. However, a designer and a builder cannot be wholly confident about the design, and therefore an appropriate factor of safety is incorporated on the design calculations.

Defects due to Inferior Workmanship

Defects due to inferior workmanship can lead to structural damage and failure. Poor workmanship is often the origin of construction defects. Even superior quality materials, if used imperfectly, may not successfully serve the planned function, or be as durable as designed. Poor workmanship is the actual cause of most construction defects. General defects due to poor workmanship are leaking roofs, cracked floor tiles, shedding paint, and other numerous problems. Proper procedures have been created for almost every construction operation, and only implementation is required. A superior quality paint that is applied to an unclean surface is likely to fail, not because the material was substandard, but because it was used with a poor quality of work.

Foundation Failure

Many building foundations are not properly designed and constructed for the existing site soil conditions. Since suitable land is often not available, buildings are constructed on soil that has inadequate bearing capacity to support the weight of the structure. Furthermore, the near surface soils may consist of expansive clays that shrink or enlarge as the moisture content is changed. Movement of foundation may occur if the clay moistening and drying is not uniform. Vegetation, inadequate drainage, plumbing leakage, and evaporation, may cause soil variation. The top soil layers provide the bearing capacity to hold the structure, and ensure stability of the foundation. If the bearing soil is inadequately compacted preceding construction, the foundation may be affected by settlement.

Types of Failures in Structures

Types of failures experienced in different structure are greatly influenced by the type of construction material used. The resisting capability of steel structure to a particular load action is not similar to that of a concrete structure for the same load conditions.

Types of Failures Experienced by Different Construction Materials

Failure caused by Steel as a Construction Material

The different features of steel material i.e. the chemical composition, heat treatments, rolling processes and other manufacture techniques employed govern the mechanical properties of steel. Steel of same grade may show different values or a result which is dependent on the rate of loading of the steel specimen, the temperature at which the test is carried out, the geometry of the specimen and the condition of the specimen.

Easy study of different failures observed in steel material can be done by classifying the same into:

- High Strength Steels.
- High strength low alloy steels.
- Tempered alloy steels.
- Quenched alloy steels.

ASTM designations are provided for each of the above-mentioned steel types. These designations are provided for light gauge cold form members, rivets, weld filler material, wire, strand, rope, cable, forgings, castings, bolts.

Common failure observed in steel rolled shapes or sections that are fabricated are:

- Failure caused in the welded areas.
- Buckling or web crippling of the sections.
- No design considerations for torsion effects result in great issues.
- Support points lack rotational restraints.
- Columns lack lateral supports.
- The compression flanges of the beams lacking lateral supports will result in the failure of the structure.
- Ignoring the effect of member stability to consider the overall stability of the structure.

Brittle fracture is a sudden failure occurring in steel structures that will result in sudden collapse of the whole system. The main causes of brittle fracture found in steel structures are:

- The loading rate on the structure is very rapid in nature.
- Corrosion of steel materials – (Mostly in the critical sections, joints, and corners where the stresses are higher).
- The loading will we provided in repeated cycles.
- The welded connections of the steel structure will be highly restrained.
- The tensile stress might be higher.
- Higher carbon content in the steel structure.
- Having flaws or notches.

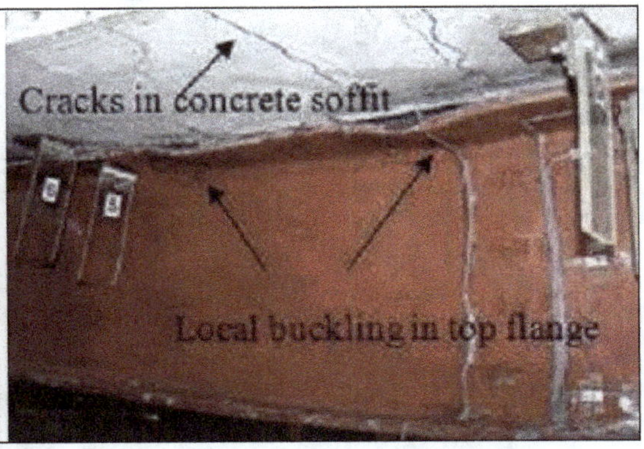

Failures caused by Concrete as a Construction Material

The concrete gain mechanical property which is dependent on the:

- Chemical Composition of the cement.
- Grading of the cement and aggregates.
- Quality of water.
- The mixing method and quality.
- Proper placing.
- Finishing of the concrete.
- Proper curing conditions.
- Small and large quality aggregates.
- The environmental conditions to which concrete is exposed.

The process of hydration bonds the various components together. This helps in increasing the mechanical properties of the concrete. More permeable the concrete more it attracts the moisture and other chemicals. This affects the internal structure of the concrete and the durability is affected extremely.

The distress in concrete are caused due to the following reasons:

- The concrete being subjected to repeated cycles of freezing and thawing.
- The concrete components subjected to extensive shrinkage and expansion.
- The chemical attack of chlorides and sulphates results in change of volume.
- Presence of high content of alumina results in the degradation of the strength.
- Chlorides penetrate the concrete that results in the corrosion of the concrete reinforcement.
- The concrete void spaces may have bacterial fermentation which results in the bursting and the disintegration.

Failures caused by Masonry as a Construction Material

Increase in the moisture content makes the masonry structure to expand and lesser moisture content makes it to contract. This is the case with brick masonry and this deformation is not reversible in nature. In the case of concrete masonry units, shrinkage is a major issue.

Failures observed in the masonry units are resulted from the following problems:

- Freezing of the water present in the joints results in the cracking.
- The metal ties and the structural steel members are subjected to corrosion.
- Cracking due to settlement issues.
- Shrinkage and bending process result in the curling of the concrete floor.
- Movement of the masonry units induces cracks extremely.
- Thermal expansion of the masonry units will result in the tension in the masonry. This will lead to cracks.
- The wall that is running in the same direction may undergo shortening and resulting in cracking of the walls.

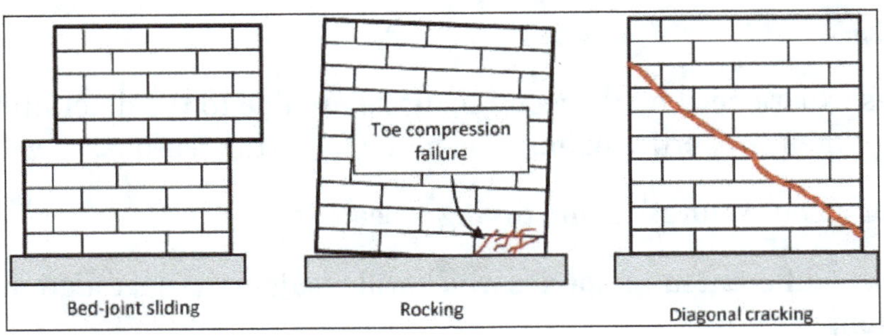

Most of the issues mentioned are due to the lack of proper consideration of design and detailing of the expansion and the control joint.

Failures in Wood as a Construction Material

Wood is a unique construction material which is mainly composed of organic cellular grained structure. Wood is hence an anisotropic material. But wood material is assumed as orthotropic material, along three principal elasticity directions along radial, tangential and longitudinal directions of the material.

The elastic properties of this material are thus studied by finding the Young's modulus. Poisson ratio and shear moduli. Most important elastic property that is used in the design is the modulus of elasticity.

Most of the failures found in this material is associated with the:

- Moisture content,
- Density,
- Splits,
- Shakes,
- Density,
- Knots,
- Checks.

Other failures experienced by the wood material are due to the deterioration and the degradation of their properties mainly due to the following reasons:

- Simultaneous saturation and drying repeatedly,
- Attack of different fungi species which will result in the degradation and decomposition,
- Termite and beetle attack,
- Wood exposure to heat and fire.

The failures observed in sizable timber structures are due to the volume changes of the wood with the varying levels of drying shrinkage, or due to inefficiency in the connection details. Improper maintenance of connections results in failure problems. Alternate wetting and drying too results in the failure.

Failure Modes of Structural Members

When a structural member is subjected to loads, its response depends not only on the type of material from which it is made but also on the environmental conditions and the manner of loading. Failure modes are different for different materials. Depending on how the member is loaded, it may fail by excessive dejection, which results in the member being unable to perform its design function; it may fail by plastic deformation (general yielding), which may cause a permanent, undesirable change in shape; it may fail because of a fracture (break), which depending on the material and the nature of loading may be of a ductile type preceded by appreciable plastic deformation or of a brittle type with little or no prior plastic deformation. Fatigue failure, which is the progressive growth of one or more cracks in a member subjected to repeated loads, often culminates in a brittle fracture type of failure.

Another manner in which a structural member may fail is by elastic or plastic instability. In this failure mode, the structural member may undergo large displacements from its design configuration when the applied load reaches a critical value, the buckling load (or instability load). This type of failure may result in excessive displacement or loss of ability (because of yielding or fracture) to carry the design load. In addition to the failure modes, a structural member may fail because of environmental corrosion (chemical action).

Following are the categories of failures modes:

- Failure by excessive deflection:
  - Elastic deflection,
  - Deflection caused by creep.
- Failure by general yielding.
- Failure by fracture:
  - Sudden fracture of brittle materials,
  - Fracture of cracked or flawed members,
  - Progressive fracture (fatigue).
- Failure by instability.

These failure modes and their associated failure criteria are most meaningful for simple structural members (e.g., tension members, columns, beams, circular cross-section torsion members). For more complicated two- and three-dimensional problems, the significance of such simple failure modes is open to question.

Many of these modes of failure for simple structural members are well known to engineers. However, under unusual conditions of load or environment, other types of failure may occur. For example, in nuclear reactor systems, cracks in pipe loops have been attributed to stress-assisted corrosion cracking, with possible side effects attributable to residual welding stresses.

The physical action in a structural member leading to failure is usually a complicated phenomenon, and in the following discussion the phenomena are necessarily oversimplified, but they nevertheless retain the essential features of the failures.

Failure by Excessive Elastic Deflection

The maximum load that may be applied to a member without causing it to cease to function properly may be limited by the permissible elastic strain or deflection of the member. Elastic deflection that may cause damage to a member can occur under these different conditions:

- Deflection under conditions of stable equilibrium, such as the stretch of a tension member, the angle of twist of a shaft, and the deflection of an end-loaded cantilever beam.
- Buckling or the rather sudden deflection associated with unstable equilibrium and often resulting in total collapse of the member. This occurs, for example, when an axial load, applied gradually to a slender column, exceeds the Euler load.
- Elastic deflections that are the amplitudes of the vibration of a member sometimes associated with failure of the member resulting from objectionable noise, shaking forces, collision of moving parts with stationary parts, etc, which result from the vibrations.

When a member fails by elastic deformation, the significant equations for design are those that relate loads and elastic deflection. For example, the elementary mechanics of materials equations, for the three members mentioned under condition (a), are $e = PL/AE$, $\Theta = TL/GJ$ and $v = PL^3/3EI$. It is noted that these equations contain the significant property of the material involved in the elastic deflection, namely, the modulus of elasticity E (sometimes called the stiffness) or the shear modulus $G = E/\left[2(1+v)\right]$,where v is Poisson's ratio.

The stresses caused by the loads are not the significant quantities; that is, the stresses do not limit the loads that can be applied to the member. In other words, if a member of given dimensions fails to perform its load-resisting function because of excessive elastic deflection, its load-carrying capacity is not increased by making the member of stronger material. As a rule, the most effective method of decreasing the deflection of a member is by changing the shape or increasing the dimensions of its cross-section, rather than by making the member of a stiffer material.

Failure by General Yielding

Another condition that may cause a member to fail is general yielding. General yielding is inelastic deformation of a considerable portion of the member, distinguishing it from localized yielding of a relatively small portion of the member.

Polycrystalline metals are composed of extremely large numbers of very small units called crystals or grains. The crystals have slip planes on which the resistance to shear stress is relatively small. Under elastic loading, before slip occurs, the crystal itself is distorted owing to stretching or compressing of the atomic bonds from their equilibrium state. If the load is removed, the crystal returns to its undistorted shape and no permanent deformation exists. When a load is applied that causes the yield strength to be reached, the crystals are again distorted but, in addition, defects in the crystal, known as dislocations, move in the slip planes by breaking and reforming atomic bonds. After removal of the load, only the distortion of the crystal (resulting from bond stretching) is recovered. The movement of the dislocations remains as permanent deformation.

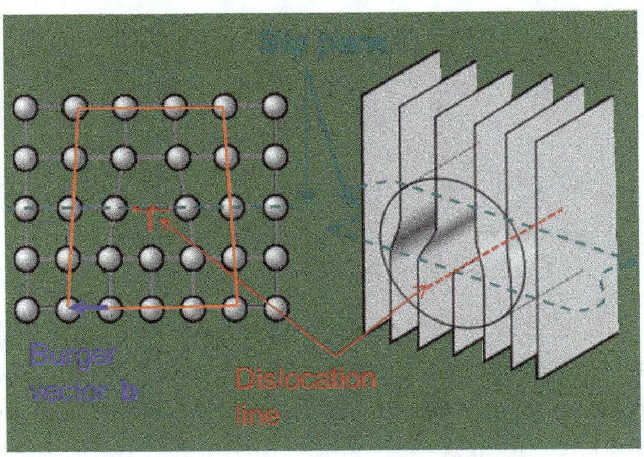

After sufficient yielding has occurred in some crystals at a given load, these crystals will not yield further without an increase in load. This is due to the formation of dislocation entanglements that make motion of the dislocations more and more difficult. A higher and higher stress will be needed to push new dislocations through these entanglements.

This increased resistance that develops after yielding is known as strain hardening or work hardening. Strain hardening is permanent. Hence, for strain-hardening metals, the plastic deformation and increase in yield strength are both retained after the load is removed.

When failure occurs by general yielding, stress concentrations usually are not significant because of the interaction and adjustments that take place between crystals in the regions of the stress concentrations. Slip in a few highly stressed crystals does not limit the general load-carrying capacity of the member but merely cause's readjustment of stresses that permit the more lightly stressed crystals to take higher stresses. The stress distribution approaches that which occurs in a member free from stress concentrations. Thus, the member as a whole acts substantially as an ideal homogeneous member, free from abrupt changes of section.

It is important to observe that, if a member that fails by yielding is replaced by one with a material of a higher yield stress, the mode of failure may change to that of elastic deflection, buckling, or excessive mechanical vibrations. Hence, the entire basis of design may be changed when conditions are altered to prevent a given mode of failure.

Failure by Fracture

Some members cease to function satisfactorily because they break (fracture) before either excessive elastic deflection or general yielding occurs.

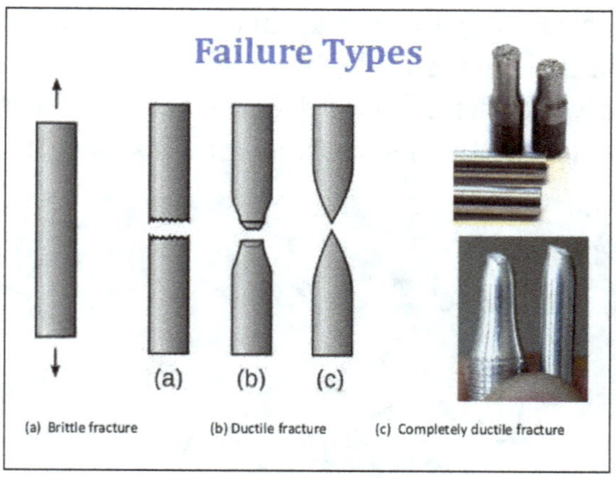

1. Sudden Fracture of Brittle Material: Some materials-so-called brittle materials-function satisfactorily in resisting loads under static conditions until the material breaks rather suddenly with little or no evidence of plastic deformation. Ordinarily, the tensile stress in members made of such materials is considered to be the significant quantity

associated with the failure, and the ultimate strength o,is taken as the measure of the maximum utilizable strength of the material.

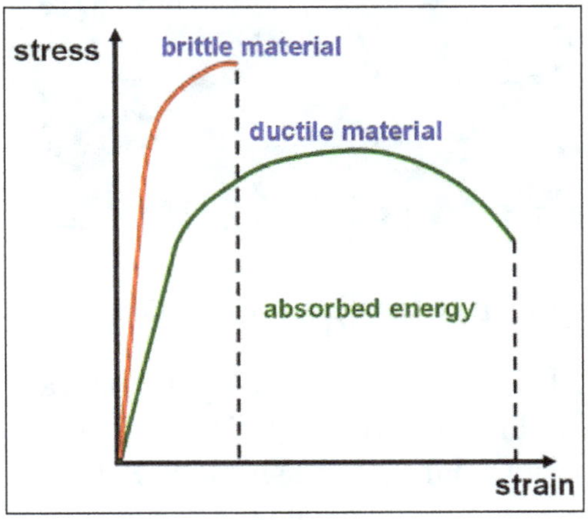

2. Fracture of Flawed Members: A member made of a ductile metal and subjected to static tensile loads will not fracture in a brittle manner as long as the member is free of flaws (cracks, notches, or other stress concentrations) and the temperature is not unusually low. However, in the presence of flaws, ductile materials may experience brittle fracture at normal temperatures. Plastic deformation may be small or non-existent even though fracture is impending. Thus, yield strength is not the critical material parameter when failure occurs by brittle fracture. Instead, notch toughness, the ability of a material to absorb energy in the presence of a notch (or sharp crack), is the parameter that governs the failure mode. Dynamic loading and low temperatures also increase the tendency of a material to fracture in a brittle manner.

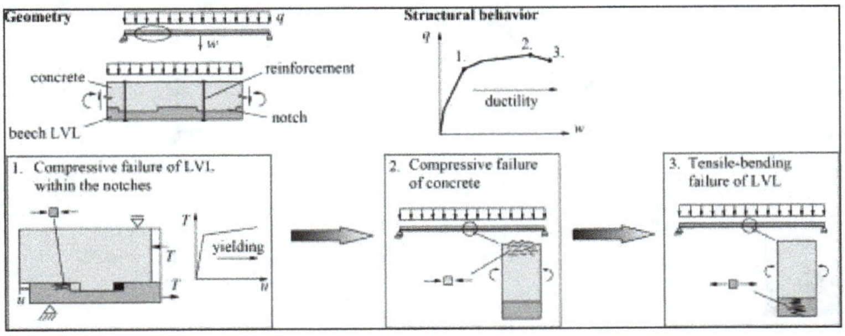

3. Progressive Fracture (Fatigue): If a metal that ordinarily fails by general yielding under a static load is subjected to repeated cycles of stress, it may fail by fracture without visual evidence of yielding, provided that the repeated stress is greater than a value called the fatigue strength. Under such conditions, minute cracks start at one or more points in the member, usually at points of high localized stress such as at abrupt changes in section, and gradually spread by fracture of the material at the edges of the cracks where the stress is highly concentrated. The progressive fracture continues until the member finally breaks. This mode of failure is usually called a fatigue failure, but it is better designated as failure by progressive fracture resulting from repeated loads.

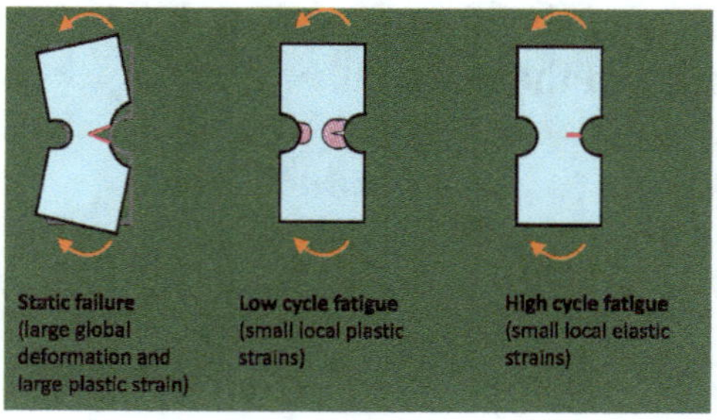

Failure by Instability (Buckling)

Some members may fail by a sudden, catastrophic, lateral deflection (instability or buckling), rather than by yielding or crushing. Consider an ideal pin-ended slender column (or strut) subjected to an axial compressive load P. Elastic buckling of the member occurs when the load P reaches a critical value:

$$P_{cr} = \pi^2 EI / L^2$$

Where, E is the modulus of elasticity, I is the moment of inertia of the cross-section, and L is the member length.

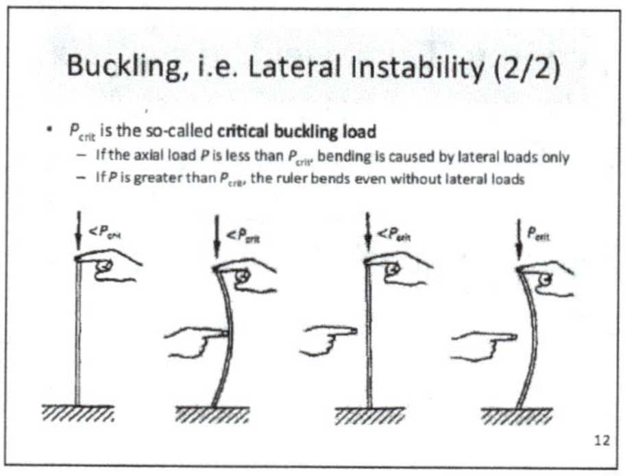

Buckling

A structure may fail to support its load when a connection snaps, or it bends until it is useless, or a member in tension either pulls apart or a crack forms that divides it, or a member in compression crushes and crumbles, or, finally, if a member in compression buckles, that is, moves laterally and shortens under a load it can no longer support. Of all of these modes of failure, buckling is probably the most common and most catastrophic.

Leonhard Euler long ago showed that there was a critical load for buckling of a slender column. A column, of course, is simply a common case of a compression member. With any smaller load, the column would remain straight and support it. With any larger load, the least disturbance would cause the column to bend sideways with an indefinitely large displacement; that is, it would buckle. The simplest case is that of a column pinned at both ends, that is, free to rotate, under a load P. Let x be the distance along the column, and y its displacement to the side. The bending moment at any point x is Py, which increases with the displacement, it should be noticed. This means that any buckling merely encourages further buckling, explaining why such failure is catastrophic.

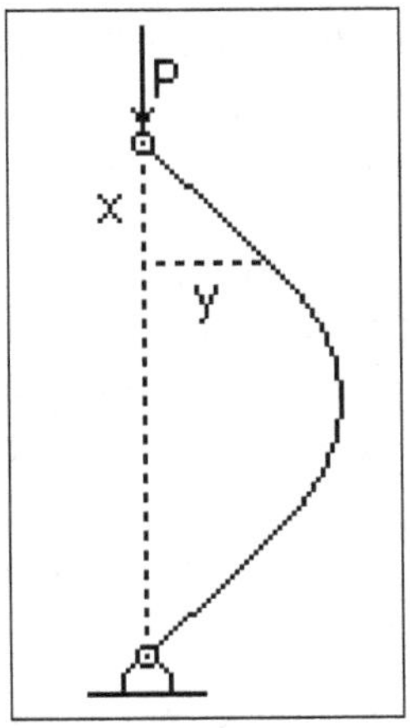

Buckling Strut.

The bending moment is related to the curvature by $Py = M = -YI\, d^2y/dx^2$, since for small y the second derivative is the reciprocal of the radius of curvature. Y is Young's modulus, and I is the moment of inertia of area of the cross-section of the column. I is the sum of elements of area times the square of their distances from the neutral axis, as in the bending of a beam. The differential equation is easy to solve, with the result that $y = A \sin ax + B \cos ax$, where $a^2 = P/YI$. $B = 0$, since $y(0) = 0$, and the condition $A \sin aL = 0$ must also be satisfied, where L is the length of the column. This means that $aL = \pi, 2\pi, \ldots, n\pi$, or the smallest value of P is given by $P' = \pi^2 YI/L^2$. The constant A itself remains indeterminate.

The interpretation of this result is that for $P < P'$ the column remains straight and $A = 0$. For $P > P'$, the column is unstable and buckles. P' is the critical load for buckling. It is found from practice that this theory gives good results for columns that are more than 30 times longer than wide. P' is proportional to $1/L^2$, so the supporting capacity of a column decreases rapidly with an increase in length. We also notice that an eccentric load will cause a decrease in strength, since there will be a bending moment and a curvature from the start. The column will be stable under sideways impacts and will return to straightness when such load is removed. The moment of inertia I will be different for different directions of buckling if the cross-section is not axially symmetrical. The direction with the minimum I will be the critical one.

There is a good incentive to look for ways to increase the critical buckling load. If the ends of the column are clamped, that is, prevented from rotating, the buckling curve changes from a half-wavelength sine to a full wavelength cosine, effectively halving the length of the column, and increasing the critical load by a factor of 4. A brace in the middle of the column that prevents it from moving to one side also halves the buckling length, and increases the critical load by a factor of 4.

For a short column, P' will become large, and at some point the compressive strength of the material will be exceeded. Let P" = Aσ be the maximum load for failure by compression. Rankine combined these results empirically to give a maximum load P by the formula $1/P = 1/P' + 1/P''$, which is quite conservative, and seems to give good results for columns between five and forty diameters long. Note that the critical load using this formula will be less than either P' or P".

If the column is made from a circular pipe, an effective unit stress for buckling can be expressed as Y/n^2, where $n = 2L/d$, where d is the diameter of the pipe. For mild steel, a safe compressive stress is about 17 000 psi, and Y = 30 000 000 psi, so P' and P" would be about equal at n = 42. A 3" diameter pipe with walls 1/4" thick and 10' 6" long could be expected to support about 40 000 lb, using a unit stress of 8 500 psi. This is a quite conservative design, with a factor of safety of at least 2 against buckling.

If the walls of the pipe are thin, another type of failure can occur, local buckling, in which the side of the pipe under compression when it deflects slightly buckles over a short distance, which is not surprising in view of the ratio of length to thickness. This was also found to occur in the upper flange of I-beams when there was too little support for a broad flange. Provision was made for this in the standard design procedure. Attempts to save material by making thin members of strong materials risks failure by local buckling.

The moment of inertia of a rectangular cross-section is $I = bh^3/12$, and for buckling b is taken as the larger of the two dimensions. For a solid circular cylinder, $I = \pi d^4/64$.

Loads

The weight or force on a joint, connection, beam, column, etc. Load is the biggest engineering factor when designing any structural element. The load can be a live load or a dead load.

The live load is the ever changing loads imposed by occupants, snow, water, ice, etc. that are constantly changing on the structure.

Dead load is the actual weight of the systems used to form the structure as well as any other type of constant loading that will be non – changing on the structure.

Special applications such as libraries introduce a heavy live load and the location of file systems cause heavy live loads on the structure. In today's technology and lack of paper files, the loading of files has been somewhat reduced. Other types of load can be imposed on a structure such as hydrostatic loads, snow loads, equipment loads, etc., each specific load must be addressed when designing any structural system.

Changing loads are most commonly the catalyst of structural problems. If a structure shifts in any way, the foundations settle elements of the framing fail, etc., and then the imposed design loads on the structure change from design to however the systems are distributing the new loading arrangement. This is why a changing snow load can cause failures or the impact of hurricane force winds which cause forces not anticipated.

Changing loads patterns will cause design issues within all structures. This is the reason that a factor of safety is included in all structural calculations. However, if the shifting is substantial enough, the factor of safety calculation may not have properly represented all the loads now imposed on the structure. Load is an important factor and an essential element in structural design.

Types of Loads on Structures

The types of loads acting on structures for buildings and other structures can be broadly classified as vertical loads, horizontal loads and longitudinal loads. The vertical loads consist of dead load, live load and impact load.

The horizontal loads comprises of wind load and earthquake load. The longitudinal loads i.e. tractive and braking forces are considered in special case of design of bridges, gantry girders etc.

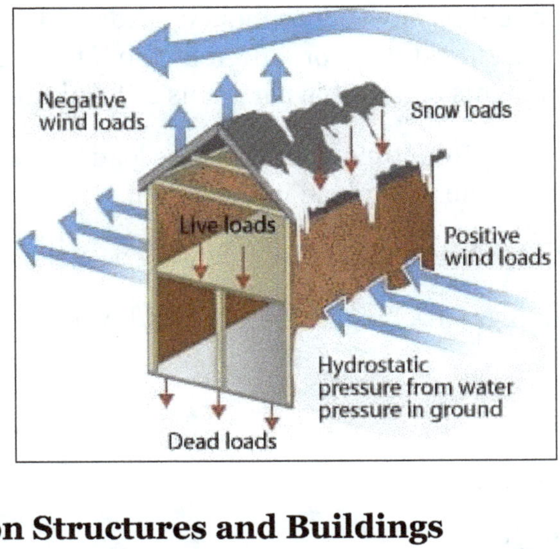

Types of Loads on Structures and Buildings

In a construction of building two major factors considered are safety and economy. If the loads are adjudged and taken higher then economy is affected. If economy is considered and loads are taken lesser then the safety is compromised.

Types of loads acting on a structure are:

Dead Loads (DL)

The first vertical load that is considered is dead load. Dead loads are permanent or stationary loads which are transferred to structure throughout the life span. Dead load is primarily due to self-weight of structural members, permanent partition walls, fixed permanent equipments and weight of different materials. It majorly consists of the weight of roofs, beams, walls and column etc. which are otherwise the permanent parts of the building.

The calculation of dead loads of each structure are calculated by the volume of each section and multiplied with the unit weight. Unit weights of some of the common materials are presented in table.

Sl. No	Material	Weight
1	Brick Masonry	18.8 kN/m^3
2	Stone Masonry	20.4-26.5 kN/m^3
3	Plain Cement Concrete	24 kN/m^3
4	Reinforced Cement Concrete	24 kN/m^3
5	Timber	5-8 kN/m^3

Imposed Loads or Live Loads (IL or LL)

The second vertical load that is considered in design of a structure is imposed loads or live loads. Live loads are either movable or moving loads without any acceleration or impact. These loads are assumed to be produced by the intended use or occupancy of the building including weights of movable partitions or furniture etc.

Live loads keep on changing from time to time. These loads are to be suitably assumed by the designer. It is one of the major load in the design.

The code gives the values of live loads for the following occupancy classification:

- Residential buildings–dwelling houses, hotels, hostels, boiler rooms and plant rooms, garages;
- Educational buildings;
- Institutional buildings;
- Assembly buildings;
- Business and office buildings;
- Mercantile buildings;

- Industrial buildings;
- Storage rooms.

The code gives uniformly distributed load as well as concentrated loads. The floor slabs have to be designed to carry either uniformly distributed loads or concentrated loads whichever produce greater stresses in the part under consideration. Since it is unlikely that any one particular time all floors will not be simultaneously carrying maximum loading, the code permits some reduction in imposed loads in designing columns, load bearing walls, piers supports and foundations.

However in multi-storeyed buildings chances of full imposed loads acting simultaneously on all floors is very rare. Hence the code makes provision for reduction of loads in designing columns, load bearing walls, their supports and foundations as shown in table.

Number of floors (including the roof) to be carried by member under consideration	Reduction in Total Distributed Imposed Loads in %
1	0
2	10
3	20
4	30
5-10	40
Over 10	50

Wind Loads

Wind load is primarily horizontal load caused by the movement of air relative to earth. Wind load is required to be considered in structural design especially when the heath of the building exceeds two times the dimensions transverse to the exposed wind surface.

For low rise building say up to four to five stories, the wind load is not critical because the moment of resistance provided by the continuity of floor system to column connection and walls provided between columns are sufficient to accommodate the effect of these forces. Further in limit state method the factor for design load is reduced to 1.2 (DL+LL+WL) when wind is considered as against the factor of 1.5(DL+LL) when wind is not considered.

The horizontal forces exerted by the components of winds are to be kept in mind while designing is the building. The calculation of wind loads depends on the two factors, namely velocity of wind and size of the building.

Designer can pick up the value of Vb depending upon the locality of the building.

To get the design wind velocity Vz the following expression shall be used:

$$V_z = k_{1.}k_2.k_{3.}V_b$$

Where,

- k_1 = Risk coefficient,
- k_2 = Coefficient based on terrain, height and structure size,
- k_3 = Topography factor.

The design wind pressure is given by:

$$p_z = 0.6\,V_z^2$$

Where, p_z is in N/m² at height Z and V_z is in m/sec. Up to a height of 30 m, the wind pressure is considered to act uniformly. Above 30 m height, the wind pressure increases.

Snow Loads (SL)

Snow loads constitute to the vertical loads in the building. But these types of loads are considered only in the snow fall places.

The minimum snow load on a roof area or any other area above ground which is subjected to snow accumulation is obtained by the expression:

$$S = \mu S_0$$

Where,

- S = Design snow load on plan area of roof,
- μ = Shape coefficient,
- S_0 = Ground snow load.

Earthquake Loads (EL)

Earthquake forces constitute to both vertical and horizontal forces on the building. The total vibration caused by earthquake may be resolved into three mutually perpendicular directions, usually taken as vertical and two horizontal directions.

The movement in vertical direction do not cause forces in superstructure to any significant extent. But the horizontal movement of the building at the time of earthquake is to be considered while designing.

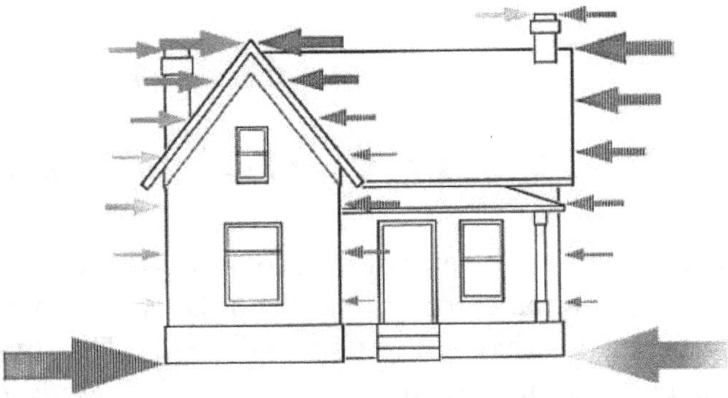

Horizontal earthquake forces (back-and-forth shaking) create 'whipping' forces in all parts of a building. These forces must transfer between parts of the building to the foundation.

Horizontal earthquake forces (back-to-back) create 'whipping' in all parts of a building. These forces must transfer between parts of the building to the foundation.

The response of the structure to the ground vibration is a function of the nature of foundation soil, size and mode of construction and the duration and intensity of ground motion. IS 1893– 2014 gives the details of such calculations for structures standing on soils which will not considerably settle or slide appreciably due to earthquake.

The seismic accelerations for the design may be arrived at from seismic coefficient, which is defined as the ratio of acceleration due to earthquake and acceleration due to gravity. For monolithic reinforced concrete structures located in the seismic zone 2, and 3 without more than 5 stories high and importance factor less than 1, the seismic forces are not critical.

References

- Building-construction-design, structural-failure-52749: brighthubengineering.com, Retrieved 21 May 2019
- Building, types-failures-construction-materials-20882: theconstructor.org, Retrieved 14 July 2019
- Failure-modes-structural-members: mechlectures.com, Retrieved 06 August 2019
- Construction-glossary: builder-questions.com, Retrieved 24 April 2019
- Structural-engg, types-of-loads-on-structure-1698: theconstructor.org, Retrieved 16 June 2019

5

Foundations in Structural Engineering

Foundations can be defined as the element that connects the structure to the ground and helps in transferring the load from the ground. Deep foundations and shallow foundations are the two types that fall under its domain. The diverse applications of foundations in structural engineering have been thoroughly discussed in this chapter.

Foundations

Foundation is the lowest part of the building or the civil structure that is in direct contact with the soil which transfers loads from the structure to the soil safely. Generally, the foundation can be classified into two, namely shallow foundation and deep foundation.

A shallow foundation transfers the load to a stratum present in a shallow depth. The deep foundation transfers the load to a deeper depth below the ground surface.

A tall building like a skyscraper or a building constructed on very weak soil requires deep foundation. If the constructed building has the plan to extend vertically in future, then a deep foundation must be suggested.

Building Foundation in Construction.

To construct a foundation, trenches are dig deeper into the soil till a hard stratum is reached. To get stronger base foundation concrete is poured into this trench. These trenches are incorporated with reinforcement cage to increase the strength of the foundation.

The projected steel rods that are projected outwards act as the bones and must be connected with the substructure above. Once the foundation has been packed correctly the construction of the building can be started.

The construction of the foundation can be done with concrete, steel, stones, bricks etc. The material and the type of foundation selected for the desired structure depends on the design loads and the type of underlying soil.

The design of the foundation must incorporate different effects of construction on the environment. For example, the digging and piling works done for deep foundation may result in adverse disturbance to the nearby soil and structural foundation. These can sometimes cause the settlement issues of the nearby structure.

Such effects have to be studied and taken care before undergoing such operations. Disposal of the waste material from the operations must be disposed properly. The construction of foundation has to be done to resist the external attack of harmful substances.

The foundation for each structure is designed such that:

- The underlying soil below the foundation structure does not undergo shear failure.
- The settlement caused during the first service load or have to be within the limit.
- Allowable bearing pressure can be defined as the pressure the soil can withstand without failure.

Purpose of Foundation

Foundations are provided for all load carrying structure for following purposes:

- Foundation is the main reason behind the stability of any structure. The stronger is the foundation, more stable is the structure.
- The proper design and construction of foundations provide a proper surface for the development of the substructure in a proper level and over a firm bed.
- Specially designed foundation helps in avoiding the lateral movements of the supporting material.
- A proper foundation distributes load on to the surface of the bed uniformly. This uniform transfer helps in avoiding unequal settlement of the building. Differential settlement is an undesirable building effect.
- The foundation serves the purpose of completely distributing the load from the structure over a large base area and then to the soil underneath. This load transferred to the soil should be within the allowable bearing capacity of the soil.

Functions of Foundation in Construction

Based on the purposes of foundation in construction, the main functions of the foundation can be enlisted as below.

- Provide overall lateral stability for the structure.
- Foundation serves the function of providing a level surface for the construction of substructure.
- Load Distribution is carried out evenly.
- The load intensity is reduced to be within the safe bearing capacity of the soil.
- The soil movement effect is resisted and prevented.
- Scouring and the undermining issues are solved by the construction of foundation.

Requirements of a Good Foundation

The design and the construction of a well-performing foundation must possess some basic requirements that must not be ignored. They are:

- The design and the construction of the foundation is done such that it can sustain as well as transmit the dead and the imposed loads to the soil. This transfer has to be carried out without resulting in any form of settlement that can result in any form of stability issues for the structure.
- Differential settlements can be avoided by having a rigid base for the foundation. These issues are more pronounced in areas where the superimposed loads are not uniform in nature.
- Based on the soil and area it is recommended to have a deeper foundation so that it can guard any form of damage or distress. These are mainly caused due to the problem of shrinkage and swelling because of temperature changes.
- The location of the foundation chosen must be an area that is not affected or influenced by future works or factors.

Shallow Foundations

Shallow foundations can be made in depths of as little as 3ft (1m), while deep foundations can be made at depths of 60 - 200ft (20 - 65m). Shallow foundations are used for small, light buildings, while deep ones are for large, heavy buildings.

Shallow foundations are also called spread footings or open footings. The 'open' refers to the fact that the foundations are made by first excavating all the earth till the bottom of the footing, and then constructing the footing. During the early stages of work, the entire footing is visible to the eye, and is therefore called an open foundation. The idea is that each footing takes the concentrated load of the column and spreads it out over a large area, so that the actual weight on the soil does not exceed the safe bearing capacity of the soil.

There are several kinds of shallow footings: individual footings, strip footings and raft foundations.

In cold climates, shallow foundations must be protected from freezing. This is because water in the soil around the foundation can freeze and expand, thereby damaging the foundation. These foundations should be built below the frost line, which is the level in the ground above which freezing occurs. If they cannot be built below the frost line, they should be protected by insulation: normally a little heat from the building will permeate into the soil and prevent freezing.

Types of Shallow Foundations

The different types of shallow foundation are:

Strip Footing

A strip footing is provided for a load-bearing wall. A strip footing is also provided for a row of columns which are so closely spaced that their spread footings overlap or nearly touch each other. In such a case, it is more economical to provide a strip footing than to provide a number of spread footings in one line. A strip footing is also known as continuous footing.

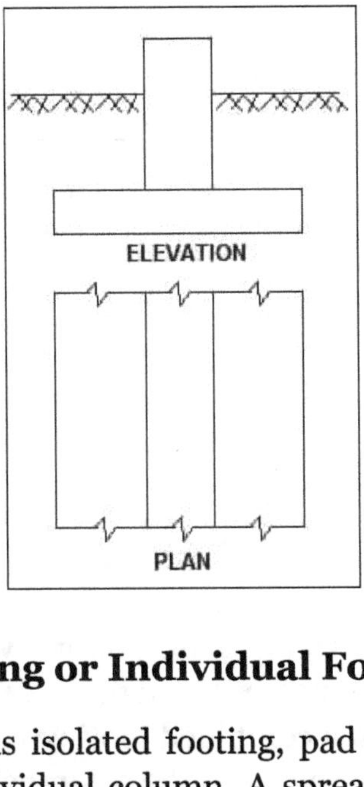

Spread or Isolated Footing or Individual Footing

A spread footing also called as isolated footing, pad footing and individual footing is provided to support an individual column. A spread footing is circular, square or rectangular slab of uniform thickness. Sometimes, it is stepped or haunched to spread the load over a large area.

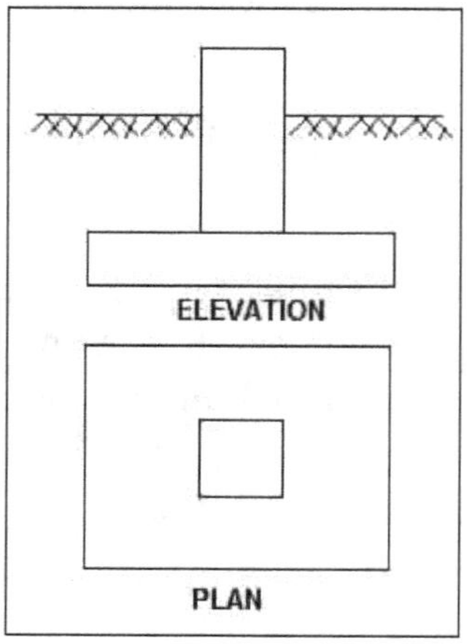

Combined Footing

A combined footing supports two columns. It is used when the two columns are so close to each other that their individual footings would overlap. A combined footing is also provided when the property line is so close to one column that a spread footing would be eccentrically loaded when kept entirely within the property line. By combining it with that of an interior column, the load is evenly distributed. A combined footing may be rectangular or trapezoidal in plan.

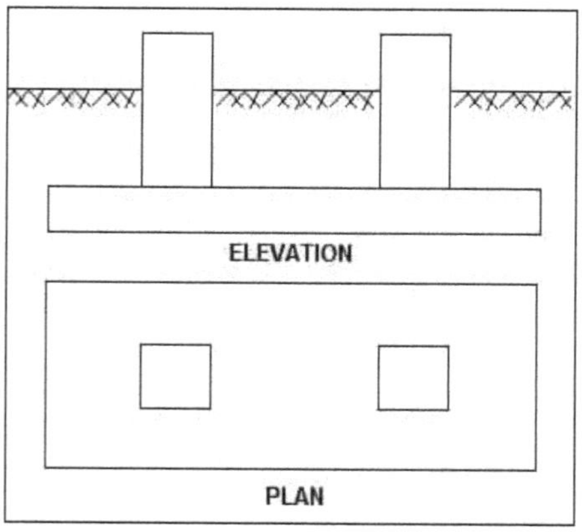

Strap or Cantilever Footing

A strap (or cantilever) footing consists of two isolated footings connected with a structural strap or a lever. The strap connects the two footings such that they behave as one unit. The strap is designed as a rigid beam. The individual footings are so designed that their combined line of action passes through the resultant of the total load. a strap footing is more economical than a combined footing when the allowable soil pressure is relatively high and the distance between the columns is large.

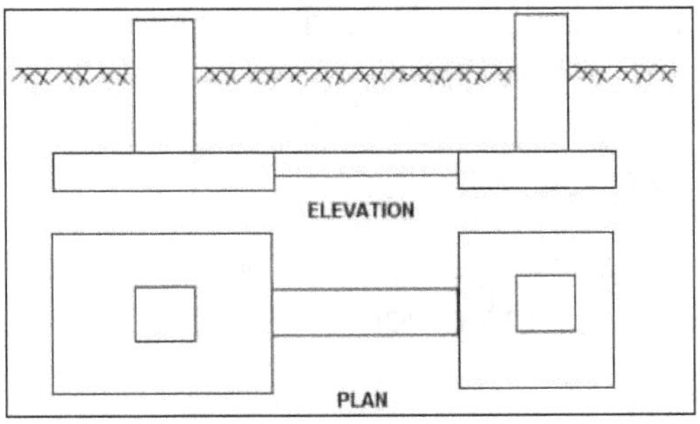

Mat or Raft Foundations

A mat or raft foundation is a large slab supporting a number of columns and walls under the entire structure or a large part of the structure. A mat is required when the allowable soil pressure is low or where the columns and walls are so close that individual footings would overlap or nearly touch each other.

Mat foundations are useful in reducing the differential settlements on non-homogeneous soils or where there is a large variation in the loads on individual columns.

Deep Foundations

Deep foundations are those founding too deeply below the finished ground surface for their base bearing capacity to be affected by surface conditions, this is usually at depths of more than 3 m below finished ground level. They include piles, piers and caissons or compensated foundations using deep basements and also deep pad or strip foundations. Deep foundations can be used to transfer the loading to deeper, more competent strata at depth if unsuitable soils are present near the surface.

Types of Deep Foundations

The types of deep foundations in general use are as follows:

Basement Foundation

These are hollow substructures designed to provide working or storage space below ground level. The structural design is governed by their functional requirements rather than from considerations of the most efficient method of resisting external earth and hydrostatic pressures. They are constructed in place in open excavations.

Buoyancy Rafts (Hollow Box Foundations)

Buoyancy rafts are hollow substructures designed to provide a buoyant or semi-buoyant substructure beneath which the net loading on the soil is reduced to the desired low intensity. Buoyancy rafts can be designed to be sunk as caissons, they can also be constructed in place in open excavations.

Caissons Foundations

Caissons are hollow substructures designed to be constructed on or near the surface and then sunk as a single unit to their required level.

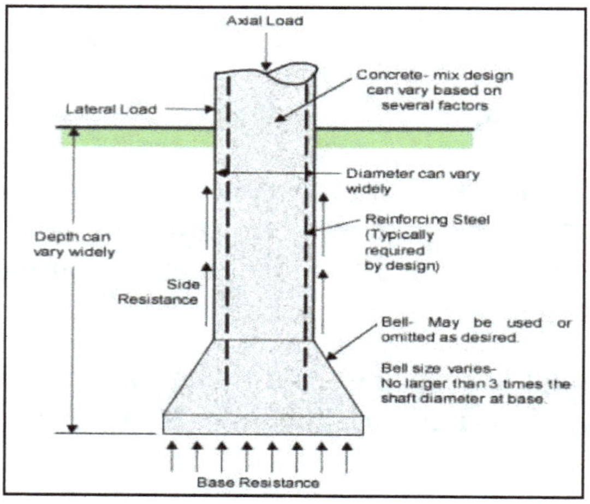

Cylinders

Cylinders are small single-cell caissons.

Drilled Shaft Foundations

Shaft foundations are constructed within deep excavations supported by lining constructed in place and subsequently filled with concrete or other pre-fabricated load-bearing units.

Pile Foundations

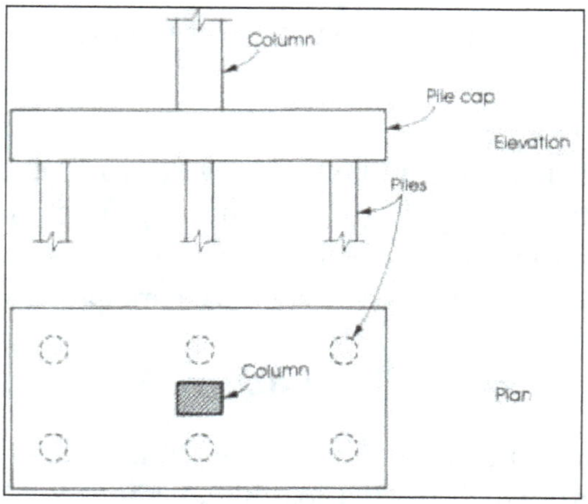

Pile foundations are relatively long and slender members constructed by driving preformed units to the desired founding level, or by driving or drilling-in tubes to the required depth – the tubes being filled with concrete before or during withdrawal or by drilling unlined or wholly or partly lined boreholes which are then filled with concrete.

Difference between Shallow and Deep Foundation

In the following table the main differences between shallow and deep foundation are given.

Sources	Shallow Foundation	Deep Foundation
The depth of foundation	The depth of shallow foundation is generally about 3 meters or the depth of foundation is less than the footing with.	Greater than shallow foundation.
Cost	Shallow foundation is cheaper.	Deep foundations are generally more expensive than shallow foundation.
Feasibility	Shallow foundations are easier to construct.	The construction process of a deep foundation is more complex.
Mechanism of load transfer	Shallow foundations transfer loads mostly by end bearing.	Deep foundations rely both on end bearing and skin friction, with few exceptions like end bearing pile.
Advantages	Construction materials are available, less labor is needed, construction procedure is simple at an affordable cost etc.	Foundation can be provided at a greater depth, Provides lateral support and resists uplift, effective when foundation at a shallow depth is not possible, can carry huge load etc.

Disadvantages	Possibility of a settlement, usually applicable for lightweight structure, weak against lateral loads etc.	More expensive, needs skilled labors, complex construction procedure, can be time-consuming and some types of deep foundations are not very flexible etc.
Types	Isolated foundation, strip foundation, mat foundation, combined foundation etc.	Pier foundation, pile foundation, caissons etc.

References

- Geotechnical,foundation-in-construction-purpose-functions-18963: theconstructor.org, Retrieved 16 June 2019
- Types-of-foundations: understandconstruction.com, Retrieved 16 May 2019
- Geotechnical, shallow-foundations-types: theconstructor.org, Retrieved 24 April 2019
- Foundations, Fountype, geocal: environment.uwe.ac.uk, Retrieved 10 April 2019
- Foundation-engineering, geotechnical-engineering, 180-difference-between-shallow-and-deep-foundation: civiltoday.com, Retrieved 30 July 2019

Permissions

All chapters in this book are published with permission under the Creative Commons Attribution Share Alike License or equivalent. Every chapter published in this book has been scrutinized by our experts. Their significance has been extensively debated. The topics covered herein carry significant information for a comprehensive understanding. They may even be implemented as practical applications or may be referred to as a beginning point for further studies.

We would like to thank the editorial team for lending their expertise to make the book truly unique. They have played a crucial role in the development of this book. Without their invaluable contributions this book wouldn't have been possible. They have made vital efforts to compile up to date information on the varied aspects of this subject to make this book a valuable addition to the collection of many professionals and students.

This book was conceptualized with the vision of imparting up-to-date and integrated information in this field. To ensure the same, a matchless editorial board was set up. Every individual on the board went through rigorous rounds of assessment to prove their worth. After which they invested a large part of their time researching and compiling the most relevant data for our readers.

The editorial board has been involved in producing this book since its inception. They have spent rigorous hours researching and exploring the diverse topics which have resulted in the successful publishing of this book. They have passed on their knowledge of decades through this book. To expedite this challenging task, the publisher supported the team at every step. A small team of assistant editors was also appointed to further simplify the editing procedure and attain best results for the readers.

Apart from the editorial board, the designing team has also invested a significant amount of their time in understanding the subject and creating the most relevant covers. They scrutinized every image to scout for the most suitable representation of the subject and create an appropriate cover for the book.

The publishing team has been an ardent support to the editorial, designing and production team. Their endless efforts to recruit the best for this project, has resulted in the accomplishment of this book. They are a veteran in the field of academics and their pool of knowledge is as vast as their experience in printing. Their expertise and guidance has proved useful at every step. Their uncompromising quality standards have made this book an exceptional effort. Their encouragement from time to time has been an inspiration for everyone.

The publisher and the editorial board hope that this book will prove to be a valuable piece of knowledge for students, practitioners and scholars across the globe.

Index

T

U

V

W

www.ingramcontent.com/pod-product-compliance
Lightning Source LLC
LaVergne TN
LVHW080608210926
848485LV00003B/121